Introduction to
BIOMES

Susan L. Woodward

D0068948

Greenwood Guides to Biomes of the World

Susan L. Woodward, General Editor

GREENWOOD PRESS

Westport, Connecticut • London

Library of Congress Cataloging-in-Publication Data

Woodward, Susan L., 1944 Jan. 20–
   Introduction to biomes / Susan L. Woodward.
     p. cm. — (Greenwood guides to biomes of the world)
   Includes bibliographical references and index.
   ISBN 978-0-313-33840-3 (set : alk. paper) — ISBN 978-0-313-33997-4
(vol. : alk. paper)
   1. Biotic communities. I. Title.
   QH541.W6334 2009
   577—dc22   2008027505

British Library Cataloguing in Publication Data is available.

Library of Congress Catalog Card Number: 2008027505
ISBN: 978-0-313-33997-4 (vol.)
      978-0-313-33840-3 (set)

First published in 2009

Greenwood Press, 88 Post Road West, Westport, CT 06881
An imprint of Greenwood Publishing Group, Inc.
www.greenwood.com

Printed in the United States of America

The paper used in this book complies with the
Permanent Paper Standard issued by the National
Information Standards Organization (Z39.48–1984).

10 9 8 7 6 5 4 3 2 1

Contents

Preface

The biome is a way to bring order to the incredible variation inherent in life on the Earth. Both similarities and differences exist in the assemblages of organisms found in different parts of the world, an observation noted by any traveler and first systematized in the eighteenth century. It is evident, at least on land, that the greatest similarity occurs in regions with the same climate. Although the actual species present are usually not the same, the structure or appearance of the vegetation is. Species adapt to their environment, and a strategy that works in one part of a climate region often works for other species in another part of the region. This basic fact is the foundation of the biome concept.

This book serves not only as an introduction to biomes, but also as the introductory volume in Greenwood Guides to Biomes of the World. With this dual purpose, it emphasizes the concept of the biome and the various environmental factors that control or determine the nature of terrestrial and aquatic biomes at a global scale. For details on one of the world's major biomes, the reader is directed to the volume dedicated to that particular biome. Information is drawn from physical geography, ecology, evolutionary biology, geology, and oceanography to explain the current patterns of life. Change is a constant factor influencing life on the Earth, so considerable attention is given to past, present, and future affects of the ever-changing global environment.

The intended audience of Greenwood Guides to Biomes of the World is the advanced middle school or high school student. The material is also appropriate for undergraduates and anyone else interested in better understanding the interrelationships of plants, animals, and microorganisms with each other and their natural environment and why species are distributed as they are around the world.

This introductory volume begins with a discussion of the development of the biome concept. A separate chapter describes other geographic trends exhibited by life, including morphological and taxonomic patterns. The third chapter focuses on basic ecological concepts that inform not only our understanding of biomes but also the vocabulary used in the study of biomes. The fourth chapter presents an overview of major environmental controls determining the distribution and nature of terrestrial biomes. The final chapter does the same for aquatic biomes. Diagrams, maps, and photographs enhance textual descriptions and explanations. Scientific terminology has been held to a minimum, but with a variety of academic disciplines represented, use of some terms unfamiliar to some readers in a broad audience was unavoidable. Therefore a fairly extensive glossary is included.

The study of biomes is important in a world facing rapid change caused by human activities on land and sea and by global climate change. If we are to conserve a major part of the Earth's biodiversity and allow the future evolution of surviving populations and species, we need to know where different species occur and understand how they came to be there, as well as how they interact with the living and nonliving elements of their environment. The study of biomes is therefore a complicated endeavor and involves delving into the deep past, the near past, and the future. A single introductory book cannot provide all the answers. Instead it may be considered successful if it leads to further questions and exploration either in other books or—better yet—in the field.

I would like to acknowledge Kevin Downing of Greenwood Press for his collaboration in developing this project and seeing it to fruition. Jeff Dixon's line drawings clarify so many points and are a major feature and enhancement of all the books in the series. Maps were skillfully prepared by Bernd Kuennecke of Radford University's Department of Geography. This volume in particular benefited from the work of all the other authors (Drs. Barbara Holzman, Bernd Kuennecke, Joyce Quinn, and Rick Roth) who wrote books for Greenwood Guides to Biomes of the World. My sincere thanks go out to all.

Blacksburg, Virginia
May 2008

1

Introduction to the Biome Concept

Anyone who has been outside in a natural setting and looked about has noticed that different sets of plants and animals live in different places. Along a stream's banks tall willows and cottonwoods grow in a narrow band. At the edge of the water, cattails may hug the shoreline. Farther inland, the riparian belt of trees gives way to different kinds of trees in the forests of humid regions or to treeless grasslands and deserts in dry areas. You might see herons or, if you are really lucky and patient, rails wading among the cattails, warblers flitting through the cottonwoods, and meadowlarks and dickcissels singing from the taller plants among the prairie grasses (if you are in North America). These predictable differences are occurring at a *local* scale and reflect ecological or habitat differences on a small section of our planet.

If you could pull back and examine the North American continent as a whole, you would notice the plant cover varies from region to region: great swaths of broadleaved forests in the east, a dark needleleaved forest in the north, open grasslands in the middle, and shrubby deserts in the southwest. At this *regional* scale you would see the structure of the plant cover but not be able to recognize the individual species comprising it. You would also see that the different climate types occurring on the continent are revealed in the different types of vegetation. Animals would not be visible from your vantage point, but they would be there with different forms characteristic in each vegetation type.

If it were possible to see the Earth as a whole, you would see that the same regional vegetation types of North America—and some new ones—occur on other continents as well. Wherever climates are similar, the vegetation looks pretty much the same. The animals, invisible at this *global* scale, may or may not be similar.

1

When you see the same type of vegetation developed under similar climates on different continents, you are looking at a biome. Study of the biome reveals just how similar or how different the component plants and animals actually are and raises questions about why plants and animals are distributed on the Earth they way they are. Answers lie in past processes and events: in Earth history (geology and climatology), evolutionary history (biology), the dispersal history of organisms (biogeography), and the current environmental conditions on the planet (geography, ecology, meteorology—and economics). This book introduces concepts from a variety of research fields that help us understand and appreciate the complexities that create and underlie each major biome.

Definition of Biome

The term biome is somewhat vaguely defined, and different interpretations of its meaning are emphasized by different fields of science and on different continents. Most often, the definition embraces the notion of a community of plants and animals characteristic of a particular climate region. In the United States, the community is often regarded as part of an ecosystem, so the definition may include not only the living parts but also nonliving or abiotic elements such as soil and the other aspects of the physical environment with which the plants and animals interact (see Table 1.1).

Table 1.1 Major Terrestrial Biomes and Their Associated Climate Types and Soils

BIOME	KOEPPEN CLIMATE TYPE	ZONAL SOIL (ORDER)
Tundra	Tundra (ET)	—
Boreal Forest (taiga)	Subarctic (Dfc, Dfd)	Spodosol
	Humid continental, warm summer (Dfb)	
Temperate Broadleaf Deciduous Forest	Humid continental, hot summer (Dfa)	Alfisol
	Humid subtropical (Cfa)	Ultisol
	Marine west coast (Cfb)	
Mediterranean Woodland and Scrub	Mediterranean (Csa, Csb)	—
Temperate Grassland	Semiarid, cold winters (BSk)	Mollisol
Tropical Savanna	Tropical wet and dry (Aw)	Oxisol
Tropical Seasonal Forests	Tropical wet and dry (Aw)	Oxisol
Tropical Rainforest	Tropical wet (Af)	Oxisol
	Tropical monsoon (Am)	
Desert	Arid, cold winters (BWk)	Aridosol
	Arid, mild winters (BWh)	

The pioneering American ecologist Frederic E. Clements coined the term in 1916 and equated it with the biotic community. In the classic work, *Bio-ecology*, co-authored by Clements and Victor E. Shelford in 1939, the biome is said to be "exemplified in the great landscape types of vegetation with their accompanying animals, such as grassland or steppe, tundra, coniferous forest, deciduous forest and the like." Clements was also the person who first developed the concept of ecological succession and the climax community (see Chapter 2), and thus he saw a biome being mapped according to the distribution of a particular climax vegetation type that reflected the regional climate. Other stages in the succession were integral parts of the total biome, so the region was actually covered by a mosaic of different vegetation types, each reflecting different stages in community development from an initial barren site to a final stable community theoretically in balance with the regional climate.

By the late-twentieth and early twenty-first centuries, American ecologists and geographers tended to define biome as a major terrestrial ecosystem, whereas European scientists continued to limit the concept to the living elements of the ecosystem, the plant and animal communities. Either way, the key ingredient and way of recognizing a particular biome is the vegetation. Vegetation refers to the growth-forms dominating or characterizing the plant cover. (The term flora is used when considering plants according to their taxonomic classifications, that is, the actual species, genera, or families found in a given area.) What is important in the biome concept is an apparent uniformity in the physiognomy of plants: their size and shape, how the foliage is arranged both vertically and horizontally, and how they respond in their life cycles to their physical environment (not only annual temperature and precipitation patterns but also edaphic conditions and disturbance factors). In plant geography, distinct physiognomic types described for large areas are known as formations, so a biome also can be defined as a major formation together with its fauna (animal life) and the environment to which they have adapted.

Scale is another factor in the definition of a biome that varies from scientist to scientist and place to place. When Clements was developing the concept, he confined his study to North America and hence his biomes were at a regional scale. This remains a useful scale for subdividing individual countries or continents into major ecosystems. A biome map of Brazil, for example, would delineate cerrado, caatinga, Amazon rainforest, Pantanal, and so forth. A global scale is usually used in textbooks in which plant formations are mapped as the same unit on every continent where they occur regardless of the local or regional differences that become apparent upon closer scrutiny. In this way, cerrado is mapped as tropical savanna and considered part of the same biome as the savannas of East Africa or Australia. Someone visiting these three parts of the biome immediately would be aware of differences in plant cover and in this example, especially, of the animals present. Yet the observer would agree that in general the vegetation has the same overall structure: a continuous groundcover of grasses with widely spaced trees or shrubs overhead. In the books comprising Greenwood Guides to Biomes of the World, biomes are separated one from the other at the global scale. The distinctions obvious at regional scales are described as regional expressions of the particular biome under consideration.

The Major Biomes of the World

General agreement exists as to how many major biomes occur on Earth's continents, and they are identified according to the dominant vegetation. Major terrestrial biomes include Tundra, Boreal Forest, Temperate Broadleaf Deciduous Forest, Mediterranean Woodland and Scrub, Temperate Grassland, Tropical Rainforest, Tropical Seasonal Forest, Tropical Savanna, and Desert (see Figure 1.1). A separate Mountain Biome is sometimes set aside for elevated regions where vegetation and animal life change with altitude, but on a world map, the zones are so narrow as to blend into each other and disappear. The Mountain Biome does not quite fit the usual definition, since it is variable over short distances and ends up including a number of plant formations adapted to sometimes different environmental conditions. The above-treeline vegetation on mountains may be thought of as a high elevation variation of the Tundra Biome, or as a separate, patchy Alpine Biome, as presented in the Greenwood Guides to Biomes of the World.

Even more of a problem to the biome concept than mountainous regions of Earth are aquatic environments, both freshwater and saltwater. In rivers and lakes and oceans, multicelled plants are largely absent, so the main criterion by which terrestrial biomes are identified is missing. Aquatic environments could be ignored

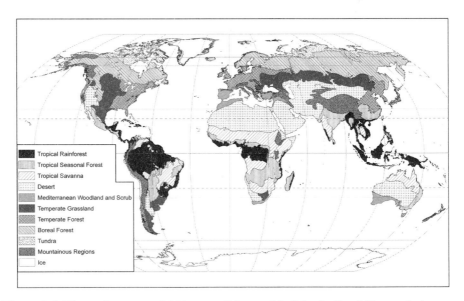

Tropical Rainforest
Tropical Seasonal Forest
Tropical Savanna
Desert
Mediterranean Woodland and Scrub
Temperate Grassland
Temperate Forest
Boreal Forest
Tundra
Mountainous Regions
Ice

Figure 1.1 The major terrestrial biomes of the world. *(Map by Bernd Kuennecke.)*

in the world biome scheme, but that would mean ignoring nearly three-fourths of the Earth's surface. Different ways of dividing freshwater and marine ecosystems into large, mappable units comparable to the terrestrial biomes have been attempted, though no general agreement has been reached among scientists. The Greenwood Guides to Biomes of the World describe three biomes in each category. For freshwater habitats, rivers, wetlands, and lakes are recognized as separate biomes even though the three are often so interconnected that it would be equally correct to view them as a single biome. For oceans, the coast, continental shelf, and deep sea are considered separate biomes. The ocean is a research frontier and, as more and more information becomes available, other ways to divide especially the open seas into biomes may gain greater validity.

Brief descriptions of each biome covered in the Greenwood Guides to Biomes of the World are provided here. Characteristic vegetation and animal life are considered in terms of their general adaptations to the respective climate or other controlling factors. The reader is directed to the respective volume for detailed discussions. The various controls or determinants of the nature and characteristics of all biomes are discussed in the remaining chapters of this book.

The Terrestrial Biomes

Tundra Biome. The simplest biome in terms of both vegetation structure and species composition of the flora and fauna is the Tundra (see Plate I). This is also the most poleward of biomes and is largely confined to the Northern Hemisphere, where it encircles the Arctic Ocean. A treeless expanse, tundra is dominated by ground-hugging vascular plants, including dwarfed shrubs and sedges. Perennial forbs dot the landscape with flowers during the short growing season. Mosses and lichens are prominent groundcovers (see Figure 1.2). Large grazing mammals are limited to two circumpolar species, the caribou or reindeer and the muskox.

Arctic hare and lemmings are abundant small mammalian herbivores. Wolves and barren-ground grizzly bears are the largest land predators. (Polar bears occupy the land when the pack ice has melted, but they are primarily marine animals preying mostly upon ringed seals. In summer, they may take reindeer and scavenge beached marine mammals.) The arctic fox is a common small predator. Its heavy dependence upon lemmings results in characteristic population cycles of three to five years, their numbers crashing when lemmings, having overgrazed the tundra plants, die off and booming again when vegetation and lemming populations recover. The Tundra Biome corresponds with polar climates in which temperatures

Figure 1.2 Profile of arctic tundra vegetation showing the ground-hugging nature of the lichens, mosses, forbs, sedges, and dwarf shrubs that make up the sparse plant cover. *(Illustration by Jeff Dixon.)*

are above freezing only 6–10 weeks of the year. The region is underlain with permafrost, which may be a greater factor in the near absence of trees than the long cold, dark winters. Thawing only a few inches below the surface in summer, tundra soils provide little anchorage for the roots of trees, so they would be easily toppled by arctic winds. Unimpeded by tall vegetation, the winds are often strong and contribute to high evaporation rates that can dessicate plants projecting more than a few inches above the ground. The thawing and freezing of the active layer above the permafrost churns the substrate and prevents the development of a true zonal soil.

Alpine Biome. Above treeline on high mountains in temperate regions of the world, vegetation resembling that of the Arctic tundra can be found, and sometimes even the very same plants live in both places. However, high-elevation communities are not exact replicas of latitudinal plant or animal associations. Mountain climates and environments vary in critical ways from polar climates. Annual and diurnal patterns of daylength and the angle of incoming solar radiation vary with latitude. The intensity of solar radiation is stronger on mountain peaks since sunlight passes through a thinner portion of the atmosphere than does that which reaches the surface at lower elevations. Temperatures are cooler at higher elevations than those considered normal for the latitude (at sea level), but they vary with the same seasonal rhythm as the surrounding lowland regions. Arctic tundra experiences months of darkness and cold, but temperate and tropical alpine areas never undergo periods of total darkness. While high elevations in the mid-latitudes may be exposed to several months with temperatures well below freezing, tropical alpine areas see temperatures fall to freezing and below every night.

Precipitation also reflects regional patterns influenced by those parts of the global circulation that affect the given latitude and then is augmented by orographic uplift or reduced by the rainshadow effect (see Chapter 3). Total annual precipitation thus may be greater than or less than regional norms.

A major control of Arctic Tundra, permafrost, is not a significant factor in mountains. Instead thin, rocky soils act to limit tree growth but provide nooks and crannies that shelter smaller plants as well as small mammals.

In the Temperate Alpine Biome (see Plate II), especially where the mountain chains run north-south (as in North America) and provide direct connections to the Arctic tundra, the same mosses, lichens, and forbs that grow in the Arctic have colonized many mountaintops (see Figure 1.3). Animal life may be quite different, however. Instead of muskoxen, for example, mountain goats and sheep browse alpine vegetation. Instead of lemmings and Arctic hare, marmots and pikas are common small herbivores.

When mountain ranges are oriented along east-west axes, as in Eurasia, or are distant from the Arctic, as in the tropics, their high elevation communities form from local floras and faunas and may bear little resemblance to the Arctic tundra. This is especially true in the Tropical Alpine Biome (see Plate II), where plants whose relatives are forbs in higher latitudes characteristically take on tree-like sizes

Figure 1.3 Profile of temperate alpine vegetation, which is similar to Arctic tundra. *(Illustration by Jeff Dixon.)*

and shapes (see Figure 1.4), as seen in the lobelias and groundsels on Mount Kilimanjaro and Mount Kenya in East Africa. Sedges and grasses may dominate large areas of tropical upland, and plants able to respond to rapid and extreme changes in temperatures on a daily basis have evolved. Among animals are a few mammals able to tolerate the low oxygen levels of the thin air, such as the vicuña of South America, and birds, such as Neotropical hummingbirds that assume a state of torpor each night, reducing energy consumption by lowering their body temperatures.

Figure 1.4 Profile of tropical alpine vegetation showing the tree-like nature of some plants. In the páramo of the northern Andes in South America, these would be puya (a bromeliad) and *frailejones* or "gray friars" (*Espeletia* sp.). In East Africa, they would be giant lobelias and giant groundsels (*Dendrosenecio* spp.). *(Illustration by Jeff Dixon.)*

Boreal Forest Biome. The boreal forest is an evergreen needleleaf forest dominated by conifers (see Figure 1.5 and Plate III). Like the tundra that lies just north of it, the taiga, as the forest is sometimes known, is restricted to the Northern Hemisphere and sweeps across both the Eurasian and North American continents in regions of subarctic and cold continental climates. Where high mountain ranges extend southward from the Canadian part of the biome, the boreal forest also extends equatorward, although in most instances different coniferous trees appear in the mountain expressions of the biome than characterize the continental forest proper. Usually, only two or three tree species dominate vast areas, so there tends to be a uniformity of color and texture across great distances.

Four kinds of needleleaved trees make up almost the entire tree flora of the boreal forest: evergreen spruce, fir, and pines and deciduous larches. The closed evergreen canopy of the spire-shaped conifers prevents much sunlight from reaching the forest floor, so forbs and shrubs are relatively rare except along forest edges. Broadleaf deciduous trees and shrubs such as alders, birches, and aspens grow in disturbed areas along stream banks or in areas recently burned. The characteristic mammals of the boreal forest proper are fur-bearers, mostly members of the weasel family such as ermine, mink, and sable, but lynx and beaver are also important. The moose (called elk in Europe) is the largest herbivore and is so closely associated with the biome that some refer to the boreal forest as the moose-spruce biome. In the western North American mountains, large mammals such as bighorn sheep, mountain goat, wapiti (elk), and mule deer browse the shrub understory. Mountain lion and wolf still prey upon the large herbivores in some places.

Evergreen plants are well adapted to short growing seasons, because they do not have to use precious time producing a new set of leaves each year. They can

Figure 1.5 Profile of boreal forest vegetation in regions dominated by spire-shaped spruces and firs. Undergrowth is sparse since little sunlight reaches the forest floor. *(Illustration by Jeff Dixon.)*

resume photosynthesis as soon as temperatures allow. In the boreal forest, where the growing season is four to six months long, the dark green pigments typical of the trees' needles absorb most wavelengths of sunlight and reradiate heat. This warms the leaf surface enough to permit some photosynthesis even before spring air temperatures might suggest it is possible. Winter's frozen soils mean a long period of drought as well as cold, so the needles held on the trees are waterproofed by waxy cuticles and sunken stomata, limiting the amount of moisture lost from the foliage.

The needleleaves of conifers decay slowly in the short cool summers of continental climate regions and produce acidic humus and soil moisture that affect soil formation. The characteristic soil-forming process is podzolization. The most prevalent soil type is the spodosol, a soil identified by its ash-colored, fine-textured A horizon. This layer has been leached of iron and aluminum compounds, as well as dark humus. Where iron oxides and humus accumulate in the B horizon, they give the soil a reddish brown hue. Acids leach most of the nutrient bases out of the soil column, and as a consequence, the forest's soils have relatively low natural fertility.

Temperate Broadleaf Deciduous Forest Biome. The vegetation of the temperate broadleaf deciduous forest (see Plate IV) evolved in response to the six- to eight-month growing season associated with warmer continental climates and humid subtropical climates on the eastern parts of continents. Winters are long enough and cold enough to impose soil moisture deficits, but the growing season is long enough to allow the annual replacement of all leaves. Dominant trees have thin, wide leaves. Usually, leaves are lobed or toothed or serrated in ways so characteristic of each species that even a young schoolchild can tell oak leaves from maple leaves or beech leaves. The deciduous habit—the shedding of all leaves for the non-growing season—eliminates problems of protecting the delicate foliage from frost and wind. The bare branches of early spring let abundant sunlight reach the forest floor and allow the development of a rich understory and herb layer. In fact, five layers of green foliage typify the summer forest: a closed a canopy of tall trees, a small tree or sapling layer, a shrub layer of deciduous and evergreen plants, an herb layer of perennial forbs or wildflowers, and a ground layer of mosses, clubmosses, and lichens (see Figure 1.6). Spring green-up proceeds from the ground up, with the wildflowers blooming and completing their annual cycles first, and the oaks, maples, beeches, and other trees of the canopy coming into leaf last.

Nut- and fruit-bearing plants are common and support a fairly rich mammal fauna, in North America characterized by tree squirrels, chipmunks, and whitetail deer as well as a number of omnivores such as raccoons and black bears. Black bears hibernate through the winter; chipmunks have interrupted phases of hibernation, storing food in their underground nests and awakening from time to time to eat. On warm days, they may be active above ground. Birds that are year-round residents are seed-eaters or omnivores. Woodpeckers are typical residents of these forests, because they can pry dormant insects from beneath the bark of trees during the winter time. All birds feed their young insects, which are extremely abundant

Figure 1.6 Profile of a temperate broadleaf deciduous forest showing the multiple layers of foliage. Since the canopy trees are the last to leaf out in the spring, sufficient sunlight exists early in the growing season to permit the development of herb and shrub layers. A ground layer of lichens and mosses forms on exposed rocks and the trunks of trees. *(Illustration by Jeff Dixon.)*

in spring and summer. In North America, a host of birds, such as thrushes and wood warblers, arrive in the spring from southern wintering grounds to nest. Once the young have fledged, these so-called Neotropical migrants return to milder climates in the southern United States, Central America, or in some cases, as far away as South America.

The broad leaves of forest trees drop off in the fall, and the thick litter that results begins to decay. The processes of decomposition halt during the winter, but are renewed with vigor in the spring just as plants are undergoing growth spurts and requiring a rich supply of nutrients. Humus derived from broadleaf trees is much less acidic than that formed under the needleleaved boreal forests and helps hold nutrient bases in the soil column and make them available to the plants. Although the major soil-forming process is podzolization, as in the boreal forest, it does not proceed to the same degree as in the northern forests. Humus, iron oxides, and aluminum oxides remain in the A horizon along with the silica compounds, coloring the horizon a grayish brown. Nutrients leached from the A horizon accumulate in the B horizon. These forest soils, particularly the group known as alfisols, were among the first exploited by early agriculturalists on all three northern continents on which the biome occurs. Until the steel plow was developed in the nineteenth century to cut the thick sod of temperate grasslands, soils beneath the temperate broadleaf deciduous forests were the most fertile zonal soils available to farming peoples.

Mediterranean Woodland and Scrub Biome. This biome occurs in a number of widely separated locations on the west sides of continents in both hemispheres between the latitudes of 30° and 40°. These rather limited regions are under the control of a mediterranean climate, one distinct from all other major climate types

Figure 1.7 Profile of mediterranean scrub vegetation. Many shrubs are evergreen and aromatic. *(Illustration by Jeff Dixon.)*

because it has a summer dry season and winter rainy season. This climate pattern means that only a short growing season is available to plants when temperatures are mild enough and soil moisture is sufficient to support photosynthesis. The characteristic vegetation in most expressions is shrubland (see Figure 1.7 and Plate V). As in the boreal forest, which also has a brief growing season, many native plants are evergreen.

To withstand the long dry summers, leaves are often reduced in size and leathery or covered with a thick waxy cuticle. Many contain aromatic oils that may defend the foliage against herbivores. Familiar seasonings such as rosemary, thyme, bay, and sage are the leaves of mediterranean plants. Fire is an important disturbance factor in the Northern Hemisphere parts of the biomes, and some of the oils are highly flammable. Shrubs such as the chamise in southern California depend on and indeed promote fire for the long-term survival of their populations on a given site. Protect the area from fire for 20 years or longer and other shrubs and trees take over; an elfin oak forest or an oak savanna may develop instead of the dense shrub cover common on the foothills around Los Angeles and other Southern California cities.

Each expression of the Mediterranean Biome has its own collection of plant species. Because all regions are isolated by mountains, deserts, and the ocean, a high degree of endemism tends to be associated with each. This pertains to plants as well as animals. Nowhere is it more evident than among the plant species of the fynbos, as the vegetation of South Africa's mediterranean region is known. Such a concentration of unique species are found in this small area that it traditionally has been designated a separate Floristic Kingdom (see Chapter 2).

Mediterranean plants have long held the interest of botanists and plant geographers because, while diverse and unique from one region to another, they display common attributes of physiognomy and have been studied as examples of convergent evolution. The picture is complicated by the fact that these areas have seen lengthy periods of human occupation, and the impact of human land uses may be such that the vegetation is not natural at all. The frequency of fire, for example, itself a natural disturbance factor, undoubtedly has been increased since human settlement. Plants that could withstand repeated burning, such as some with rosette growthforms or with thick, corky bark, were selected. Clearing for agricultural lands and forest products took place in the Mediterranean proper since at least the days of Homer. The result was soil erosion, increased soil aridity, and thin stony substrates. Plants that could withstand dry thin soils were more successful than those that required deeper richer soils. Grazing livestock such as goats and sheep created selection forces favoring plants armed with unpalatable oils and thorns. This may be a biome largely formed by human activities.

In contrast to the plant life, fewer animals are restricted to this biome, although some endemic birds and reptiles are known. Soils also seem not to be determined by climate or vegetation and fit no zonal scheme. Bedrock type and geological process are apparently more important in the formation of rendzinas and other mediterranean soil types.

Temperate Grassland Biome. Associated with semiarid continental climates in the mid-latitudes, temperate grasslands are dominated by perennial grasses and forbs (see Figure 1.8 and Plate VI) well adapted to marked seasonal temperature changes and 10–20 in (250–500 mm) of rainfall each year. Both the grasses and forbs die back at the end of the growing season. Their growth renewal buds lie at or just beneath the ground surface, a location that protects them not only from cold but also from fire and grazing, both major disturbance factors in the biome.

Grasses' tillers or narrow, upright stems reduce heat-gain in the hot summers. An intricate and deep root system traps moisture and nutrients. Grass may be either turf- or sod-forming types or bunchgrasses. The former regenerate by means of rhizomes or underground stems along which new plants spring forth. The latter grow as unconnected clumps and reproduce by seed. Tall sod-forming grasses are more likely found in moister parts of the grassland biome; shorter bunchgrasses become dominant in drier regions.

The temperate grassland fauna is very low in diversity, especially when compared with the tropical grasslands or savannas of Africa. In North America, the characteristic large mammals are bison and pronghorn, the sole member of the endemic Nearctic family, Antilocapridae. Among the more common and visible small mammals are the prairie dog and some ground squirrels. Carnivores include coyote (actually an omnivore), badger, and the federally endangered black-footed ferret.

Figure 1.8 Profile of a temperate grassland showing the abundance of perennial forbs that grows among the grasses. In spring and summer, the steppes and prairies assume different hues depending on which plants are in flower. *(Illustration by Jeff Dixon.)*

The last two are members of the weasel family. On the Russian and Ukrainian steppes, the fauna formerly included wisent, tarpan or wild horse, and saiga antelope, but today these are either extinct or rare. Mole rats are conspicuous by virtue of their many mounds. Polecats and other members of the weasel family are among the larger, extant carnivores.

Calcification is the dominant soil-forming process in semiarid regions. Mild leaching and the annual diedown of the aboveground parts of forbs and grasses result in a high organic content. Capillary action moves dissolved compounds upward in the soil column, and when the soil moisture evaporates, the minerals precipitate out. Concentrations of calcium carbonate in the B horizon typify the dark brown mollisols that develop under temperate grasslands. When this process works on loess that itself is rich in calcium, the world's most fertile soils are created, the chernozems (a Russian term meaning black soil). Loess and hence chernozem underlie the eastern prairies of the United States, the pampas of South America, and the steppes of Ukraine and Russia. Once the development of the steel plow enabled breaking through the thick sod, most temperate grasslands became the so-called breadbaskets of the countries owning them; and wheat and other cereals replaced the native grasses.

Tropical Rainforest Biome. Tropical rainforests (located generally between 10° N and 10° S) are Earth's most complex biome in terms of both vegetation structure and species richness. Life abounds in fantastic array in an equatorial climate optimal for plant growth (see Plate VII). Year-round temperatures are warm and moisture abundant. The real limiting factor is sunlight itself. The trees of the rainforest are typically broadleaf evergreens. The dense foliage casts a dark shadow on the forest floor. The struggle for light has influenced the structure of the vegetation and the variety of growthforms associated with it (see Figure 1.9).

Rainforest trees are arranged, more or less, in three layers, known as the A (above canopy or emergent) layer, B layer (the closed canopy or top of the forest), and C layer, another dense layer of tree crowns below the main canopy. Sunlight is readily available on the top of the B layer, but greatly reduced below it where air movement is also greatly reduced. Daily and seasonal drought may affect crowns of A layer trees, but in the C layer and below it is always humid. Below the trees is a shrub and sapling layer in which young plants delay their growth, waiting for a break to occur in the canopy due to treefall, at which time sunlight pours in and they undergo a spurt of growth that quickly fills the gap.

The ground layer is sparse due to low light and, perhaps surprisingly, low rainfall. The canopy above not only intercepts incoming sunlight but also prevents one-third of the rainfall from reaching the forest floor. To cope with such situations, small plants tend to grow not on the floor but high in the trees. A commonly occurring growthform is that of the epiphyte or air plant. These plants perch on the branches of trees in the A and B layers, their roots dangling in the air. They obtain moisture and nutrients from the atmosphere, rainfall, and leaf fall through the

A-layer

B-layer

C-layer

Sapling
layer

120 ft

90 ft

60 ft

Figure 1.9 Profile of a tropical rainforest. Three tree layers make this the most structurally complex biome. Smaller plants have evolved ways to position their foliage in the upper layers of the canopy where sunlight is plentiful. Among the growthforms that accomplish this are epiphytes, stranglers, and lianas. The sparse cover of plants on the forest floor consists of species adapted to low light levels either by being parasites acquiring nutrients from the roots and stems of larger plants or by suspending growth until a gap occurs in the canopy above. *(Illustration by Jeff Dixon.)*

canopy, not from their host plant or from the soil. Many ferns, orchids, bromeliads, and even cacti (the latter two occur only in the Americas) have adopted this growthform. Another strategy employed by certain groups of plants is to be rooted in the soil but have the photosynthetically active and reproductive parts of the plant—leaves and flowers—high in the tree canopy. Stranglers accomplish this by starting out as epiphytes. The young plants extend their roots down the host, wrapping it in a net of strong woody roots. Lianas or woody vines do the opposite. They germinate in the soil and extend their stems up the trunks and along the branches of host trees whenever a gap in the canopy lets in sunlight until their leaves and blossoms reach the top of the forest. Climbers are green-stemmed vines with large leaves to collect as much light as possible in the dark lower part of the rainforest. A final group of plants does not depend on photosynthesis at all for energy or nutrients. Among these so-called heterotrophs are parasites that tap into the roots of trees or stems of lianas and steal the energy-rich nutrients they produce and saprophytes that derive their nutrients from decaying organic material.

Unlike the broadleaf trees of the temperate forests, tropical trees tend to have entire leaves that lack lobes and other indentations and look pretty much the same from species to species—a real problem when you consider that there may be 40 to 100 different kinds of tree in every 2.5 ac (1 ha). Correct identification requires the study of flowers and fruits high above the forest floor or of the DNA in a laboratory. Not surprisingly many have yet to be described by scientists.

Animal life in the rainforest biome is highly diverse. Most forms are adapted to life in the trees. They run, jump, or swing from branch to branch. With a yearlong growing season and plants not following a strict seasonal schedule of flowering and fruiting, fruits are available all year. This is the only biome in which animals can have specialized diets based on fruits. Bright colors, distinct patterns, and loud voices are other common characteristic of animals living in dense growth and among many similar species. Rainforests are famous for their brilliant butterflies and birds, colorful tree frogs, and loud monkeys and apes.

Precipitation is often in excess of 100 in (2,500 mm) a year in tropical rainforest climates. Combined with the fact that much of the biome lies on ancient bedrock, the foundation blocks of the respective continents, most nutrients were leached away long ago. The dominant soil-forming process is laterization, in which abundant moisture has mobilized the silica compounds of the A horizon and washed them out, leaving behind iron and aluminum oxides. Iron and water combine to form red rust, and tropical soils (oxisols) are typically bright red. Sometimes the aluminum oxides occur in such high concentrations that the soil is mined as a source of aluminum metal. The paradox of nutrient-poor soils and luxurious vegetation is possible because warm temperatures allow rapid decomposition of litter and surface roots quickly absorb nutrients released through decay. The ecosystem as a whole is not lacking in nutrients. It is just that, at any given moment, the vast majority of nutrients are held in the living plants and animals and not in the soil.

Tropical Seasonal Forest Biome. An often-neglected biome but one important in terms of large mammal diversity and human alteration is that of seasonal or dry tropical forests (see Plate VIII). Simpler in structure than the rainforests, these forests usually have only one or two tree layers. Some or all of the trees are deciduous part of the year during the dry season. The longer the dry season the greater the proportion of deciduous broadleaf trees and the shorter the height of the trees, the smaller their leaves, and the greater the likelihood plants will possess thorns. The biome is complex in that it includes tall forests bordering the rainforests and grades into low thornscrub bordering the hot deserts of the world. The general sequence of vegetation from about 10° to 25° of latitude is semievergreen forest; deciduous seasonal or dry forest; and thorn forest or thornscrub (see Figure 1.10). The biome is associated with tropical wet and dry and tropical monsoon climates. The former also is associated with tropical savannas and the latter with tropical rainforests. In some locations, especially in Asia, tropical savannas are quite likely derived from tropical seasonal forests heavily affected by human-set fires and livestock grazing. Moister parts of the biome grow on typical tropical oxisols.

Tropical Savanna Biome. In vegetation studies, the term savanna designates a vegetation characterized by a continuous groundcover of grasses with an open canopy of shrubs or trees above it (see Figure 1.11 and Plates IX and X). Some savannas are light woodlands with closely spaced trees that still let enough sunlight through

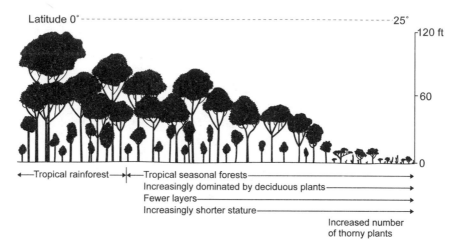

Figure 1.10 Latitudinal transect from the edge of equatorial rainforests to the equator-ward margins of warm deserts. As distance from the Equator increases, the trees become shorter and more species are deciduous, dropping their leaves during the dry season. The structure of the vegetation also becomes simpler as fewer layers develop. In thorn forests and thornscrub, the leaves of many woody plants are small and thorns are prevalent. *(Illustration by Jeff Dixon.)*

for grasses to flourish. Others are nearly pure grasslands, with only a few scattered trees or shrubs. Most are somewhere in between. Savannas develop in areas with tropical wet and dry climates.

At least five months of the year see less than 5 in (125 mm) of rain. Total annual precipitation averages between 30 and 50 in (760 and 1,270 mm). In general, grasses are better able to withstand prolonged drought than are woody plants, and so they predominate. Still, certain drought-adapted or drought-resistant trees and shrubs may gain a foothold. However, trees seem to be held in check by frequent

Figure 1.11 Profile of a typical African savanna with tall grasses completely covering the ground and flat-topped acacias and other trees forming an open canopy above them. Many types of savanna are recognized and identified either by the spacing of the woody plants or by the types of woody plants that make up the upper canopy layer. *(Illustration by Jeff Dixon.)*

fires, poor drainage (waterlogged conditions), excessive drainage (droughty conditions, as can develop in sandy substrates), or strong browsing pressures from large mammals. The biome is often viewed as being more under the control of these edaphic and disturbance factors than climate.

The world's greatest variety (more than 40 different species) of ungulates (hoofed mammals) is found on the savannas of Africa. The antelopes are especially diverse and include eland, impalas, gazelles, oryx, gerenuk, and kudu. Buffalo, wildebeest, plains zebra, rhinos, giraffes, elephants, and warthogs are among other herbivores of the African savanna. Up to 16 grazing and browsing species may coexist in the same area. They divide the resources spatially and temporally; each having its own food preferences, grazing or browsing height, time of day or year to use a given area, and different dry season refuges. Most gather in herds, and some migrate seasonally from one feeding ground to another, as famously demonstrated by the zebras and wildebeests of the Serengeti plains. The abundance and diversity of herbivores support many different carnivores including large and small cats, hyenas, jackals, and wild dogs.

The large mammals of East Africa's savannas are what one usually thinks of when one hears the word savanna. But though the image is well burned into our minds, other savannas lack such a large complement of ungulates. In South America, few animals are restricted to the savanna biome; most savanna mammals and birds are also found in other biomes on the continent, either the tropical rainforests or tropical seasonal forests.

Termites are especially abundant in the tropical savannas of the world, and their tall mounds or termitarias are conspicuous elements of the savanna landscape. These detrivores are important in soil-formation; their termitaria provide shelter for other animals; and the termites themselves are the beginning of food chains that include the strange-looking anteaters of the Neotropics and aardvarks and pangolins of Africa.

Most tropical savannas occupy old geologic surfaces. Sufficient time and moisture have existed for the development of oxisols. A few savannas, however, such as those developed on younger volcanic substrates, may be associated with relatively nutrient-rich soils.

Desert Biome. The desert biome contains many plants and animals that have evolved a variety of means for tolerating or avoiding the extreme aridity and high temperatures typical of the desert environment. Rarely, if ever, are desert areas devoid of life. Climatically, deserts are usually defined as regions receiving less than 10 in (250 mm) of precipitation a year. A number of geographic conditions may create such aridity, among them a position on the west coast of continents between 20° and 30° latitude, where cold currents flow offshore; a location on the leeside of major mountain chains; a position in the interior of large landmasses; and an association with the major subtropical high-pressure zones. Low-latitude deserts are classified as warm deserts (see Plate XI) because temperatures rarely go

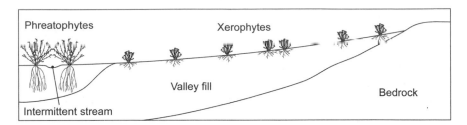

Figure 1.12 Profile of warm desertscrub. Deserts are dominated by widely spaced shrubs, either evergreen or deciduous. Succulents and perennial forbs are frequently present. After rare rains, a carpet of ephemerals suddenly appears and adorns the desert floor with bright colors; but most of the time, the wildflowers are invisible and exist as seeds. *(Illustration by Jeff Dixon.)*

below freezing. Interior deserts in the middle latitudes (see Plate XII) experience continental-type temperature extremes, and temperatures below freezing are common and often prolonged features of the winter season. These deserts are classified as cold deserts.

Shrubs are the dominant growthform of the world's arid regions (see Figure 1.12). They may be evergreen or deciduous, typically have small leaves, and frequently have spines or thorns or contain aromatic oils. Shallow but extensive root systems capture rainwater before it has a chance to evaporate or percolate into the ground. Such shrubs are well adapted to tolerate extreme drought and called xerophytes. They are widely spaced across the desert floor and, except after rains when short-lived annuals (ephemerals) may cover the desert floor, the ground between them is bare of vegetative growth.

Other growthforms are adapted to reach water available at considerable depth beneath the surface or to store water in their tissues, so in a sense, they do not truly live in a water-deprived environment. Long tap roots enable plants called phreatophytes to tap into groundwater, which may lie fairly close to the surface near seasonally flowing streams. Succulents store water in their tissues. Most familiar are the many cacti of the New World deserts, but many other plants families around the world have members able to hold water drawn up from the roots after a rain in their stems, leaves, or special underground structures. A fascinating variety of sometimes bizarre succulent plants have evolved in the world's warm deserts and west coast fog deserts. Perennial forbs are often abundant. They store moisture and nutrients in bulbs and or rhizomes. The aboveground part of the plant is often entirely lacking much of the year, when the plant is dormant; however, after sufficient rain, leaves sprout rapidly and the plant flowers. Ephemerals—annual forbs and grasses that complete their life cycle in only a few weeks—are abundant but may lie hidden in the soil as waterproofed seeds. Only after enough rain has fallen to dissolve the protective seed coating do the seeds germinate and for a few short weeks carpet the desert with flowers. This does not necessarily happen every year, but when it does, it produces one of nature's spectacles.

Animals of the desert biome must also possess ways to cope with a shortage of free water. Behavioral adaptations such as being active only at night or at dawn and dusk (that is, being crepuscular) lets animals avoid the hottest and driest part of each day. Some simply shelter themselves in the shade of a bush or rock outcrop during the heat of day. Others retreat below ground where temperatures are lower and humidity higher. The size, shape, and color of desert vertebrates are often distinct from close relatives inhabiting cooler or more humid climate regions. Such morphological adaptations include smaller body sizes and longer appendages (ears, legs, wings, dewlaps, even snouts) that let warm-blooded animals radiate excess body heat to the atmosphere. Skin, fur, and feathers may be lighter colored to reflect sunlight and help prevent the absorption of heat from the environment. Rarer, but equally important, are physiological or metabolic adaptations such as aestivation (dormancy during summer), the absence of sweat glands, the concentration of urine, localized deposits of fat in tails or humps, and specialized salt glands that secrete salt without the loss of precious body fluids.

Reptiles are exceptionally well adapted to drylands and, not surprisingly, diverse in deserts. Lizards, snakes, and tortoises have waterproof skin and produce uric acid instead of water-wasting urine. Their eggs have hard shells that protect the developing embryos and keep them from drying up. Cold-blooded reptiles regulate their body temperatures by moving onto sunny sites early in the day to absorb sunlight and warm up and then retreating to shade or underground when they need to prevent overheating.

Birds can fly to streams or pools of water if available in nearby foothills or at springs, and so many do not exhibit special adaptations to desert living. In Australia, where waterless areas are expansive, some bird populations do not maintain regular breeding seasons but nest only when sporadic rainfall ensures an adequate supply of insects and seeds to feed their nestlings and fledglings.

Calcification is the dominant soil-forming process in arid lands. At best, the development of horizons is poor, and calcium carbonate accumulates at or near the surface. The sparse plant cover and tiny leaves of most shrubs produce little humus. Soils, mostly aridosols, typically have a pale gray color.

The Freshwater Aquatic Biomes

River Biome. The organisms of the river biome are adapted to life in flowing water. Rivers occur in every geographic region (two are known from Antarctica), but habitat characteristics differ from one climate region to another. Everywhere, rivers are inseparable from the lands through which they flow and from which surface runoff drains into them, their watersheds. Together the stream network and the watershed are constantly reshaped and altered by geomorphic processes of erosion, sediment transfer, and sediment deposition. The river channel itself changes with age and with distance from the source, as streams downcut and widen their channels. A variety of habitats ensue from the stream banks and floodplains, to

Figure 1.13 This low-order stream in Virginia demonstrates the variety of freshwater habitats typical of streams. Deeper pools occur beneath undercut banks on the outside of meanders; downstream between the bends in the stream are riffles; and on the inside of meanders are small gravel bars. *(Photo by author.)*

waterfalls or rapids and pools, to riffles and pools, to meanders with undercut banks and deep pools on outer bends and shallow point bars of gravels and sands on inner bends (see Figure 1.13 and Plate XIII). Most rivers undergo annual changes in water level with floods in rainy seasons or when snow melts in the upland regions of their headwaters and low water during dry seasons or hot summers when evaporation rates are highest. During low-water periods, the freshwater habitat that extends from the river into the hyporheic zone—the sides and bottom of the channel where water is exchanged between stream and groundwater—can become an important sanctuary for small bottom-dwelling (benthic) organisms.

Water flowing downstream exerts drag on anything in the moving water, so organisms inhabiting a stream face the challenge of being swept away. Body shapes are often flattened or streamlined to let water pass smoothly over or by them. Some attach themselves to streambed sands and rocks with adhesives or silken threads. Others have claws with which to clasp onto rocks or suction-cups that hold them in place. Avoidance of flood waters when flow velocity increases is possible for those animals that are able to move to the more-sheltered areas in the stream channel or bury themselves in the channel bottom, or that are able to leave the stream altogether.

Plants as well as animals must be adapted to changing flow conditions. Few large plants are able to do this. Trees such as willows and cottonwoods are among those that thrive in riparian situations. Their stems are flexible enough to withstand large volumes of water passing by them and the thin leaves of willows reduce drag when their branches become submerged during floods. Most other large plants associated with streams are either loosely rooted or not rooted in the stream bed at all. They normally have leaves that float on top of the water and live in calmer backwaters or are floating mats that ride the river downstream.

Most of the photosynthesis that goes on in streams is associated with the floating single-celled organisms (algae and cyanobacteria) of the plankton and the algae, mostly diatoms, that make up the periphyton, the biofilm that coats the submerged surfaces of substrate, plants, and animals. Considerable energy, however, derives from organic materials washed into the stream from the land. Dead leaves and other plant parts; pollen, fruits, and seeds; and dead animals and animal wastes feed a decomposer community of bacteria. Protozoans and small invertebrates feed on the bacteria, transferring energy and nutrients through a microbial food web. These tiny animals themselves are food for larger organisms.

The macroinvertebrates and fish in stream ecosystems are classified into functional feeding groups or guilds by ecologists. Grazers, such as snails and water pennies, scrape away the periphyton but also may consume microinvertebrates, bacteria, and organic detritus that coat underwater surfaces. Shredders feed on coarse particulate organic matter (CPOM), mostly leaf debris that has been in the water long enough for bacteria and fungi to colonize it. Among the shredders that break down leaves into smaller particles are the larvae and nymphs of caddisflies, stoneflies, scuds, and some true flies, and beetles. The organisms that filter fine particulate organic matter (FPOM) from the water, as mussels do, or from the bottom sediments, as oligochaete worms do, are known as collectors. They either spin nets or extend net-like body parts into the current. The last guild is composed of animals that prey upon other animals, the predators. In this group are invertebrates with mouthparts adapted for piercing or biting, such as hellgrammites, the larvae of dragonflies and many kinds of dobsonflies, stoneflies, caddisflies, and beetles. Amphibians and fishes are also in this guild. Fishes exhibit a variety of body shapes, specialized body parts, and behaviors that reflect their feeding preferences and strategies. Many animals change feeding guilds at different stages of their life histories. They may be plankton-feeders when young and become predators as adults.

One important concept that tries to bring order to the great diversity of riverine habitats and conditions found around the world is the River Continuum Concept (RCC). This model generalizes conditions along the length of a river from its small headwater streams to the broad mainstem near the river's mouth. Headwater streams are steep and fast-flowing; the channel is narrow and usually shaded by overhanging vegetation. Typically large inputs of terrestrial production provide much of the energy and nutrients for stream-dwelling organisms. In these upper parts of the stream system, a large percentage of the invertebrates are shredders. The few grazers indicate the paucity of the periphyton. The fish community consists of cold-water species with high oxygen demands.

Mid-size streams in the system are wider, so more water is exposed to sunlight. Photosynthesis among the plankton and periphyton becomes of major importance at the beginning of food chains. The primary production of microscopic plants is augmented by that of macrophytes. Where sediments accumulate in calmer stretches of the river, large, rooted plants such as cattail and floating plants such as pond lilies become established. With less of the channel covered by the canopy of riparian trees, leaf fall and other terrestrial inputs are reduced. Still, organic detritus flows into these parts of the system from upstream.

In large rivers, even though overhanging vegetation is a minor factor, increased sediment loads cloud the water and reduce the amount of sunlight available to the plankton and periphyton. Very deep water can have the same consequence. A dependency on energy flowing in from upstream as FPOM is characteristic. The sediments are rich in organic matter. Collectors such as mussels and worms dominate the invertebrate community. Warm-water species dominate the fish community.

Wetlands Biome. Wetlands can be found in every climatic situation; they are absent only from Antarctica. Globally, they form under a variety of conditions determined by geomorphic, hydrologic, and biological factors. Many are directly connected to rivers or lakes, but others are totally separate entities.

Wetlands are the one freshwater biome in which large vascular, water-loving plants are conspicuous. They help to define different types of wetland. Emergent grasses and grass-like reeds and cattails dominate in marshes. Woody plants are the dominants in swamps. Sphagnum mosses are indicators of bogs and fens (see Plate XIII).

Two other elements that define wetlands are soils saturated or waterlogged all or part of the year and a water table close to or even above the surface that causes ponding of water at least seasonally. The pattern of rise and fall of water levels (the hydroperiod) is a significant determinant of wetland characteristics. Wetlands are divided into tidal and nontidal types. Tidal wetlands may be flooded only during spring high tides or during storms; regularly flooded and exposed on a daily basis; exposed only irregularly during spring low tides; or never exposed, even at low tide. Among the nontidal varieties are permanent, intermittent, seasonal, and temporary wetlands.

Some wetlands do form as the result of local rainfall and are not the product of high water tables. These so-called ombrotrophic wetlands are rare and develop only on substrates such as clays or laterites and other hardpans that prevent the downward percolation of water.

Living in a wetland environment requires adaptations to low or no oxygen, fluctuating water levels, and often an accumulation of toxic compounds, such as hydrogen sulfide, produced by anaerobic decomposers (bacteria). Some one-celled organisms have a cellular biochemistry that allows them to function without oxygen. Higher plants may have special tissues (aerenchyma) and structures (pneumatophores) through which they can obtain oxygen and direct it to their roots, which are anchored in the saturated, oxygen-depleted soil.

Wetland vegetation tends to be arranged in concentric zones reflecting a gradient of increasing soil wetness and running from dry upland soils through zones of increasingly lengthy periods of saturation to the innermost areas that are either permanently flooded or open water. Trees tolerant of saturated soils form the outer edges. These may be willows or swamp maples or, in bog situations, black spruce and larch. Emergent cattails and sedges grow in the next ring on periodically flooded terrain. Shallow areas that are always inundated are habitat for rooted plants with leaves that float on the surface, such as pond lilies and spatterdock. Once the open water becomes too deep for such floating plants to reach the surface, submerged aquatic vegetation such as pondweeds and milfoils flourish. Rooted plants disappear altogether in water too deep for sunlight to penetrate.

In wetlands of all types insects are common. Many are also common in rivers and lakes. Their larvae are important constituents of detritus food chains and provide food for fish, amphibians, and birds. Insects widely distributed geographically and occurring in all types of wetlands include mosquitoes (order Diptera) and dragonflies and damselflies (order Odonata).

Frogs, toads, and salamanders all inhabit wetlands. Some may be present only in their larval stages as tadpoles and newts; but many frogs and salamanders are lifelong residents. While few reptiles are full-time inhabitants of water, many are closely linked to wetland habitats. Turtles and members of the crocodile order, including the American alligator, must live near water.

Fishes are present in many wetlands, but unless deliberately introduced, they are absent from prairie potholes and other wetlands not connected to a river or lake. Few occupy the acidic waters of bogs, although the world's smallest fish was recently discovered in such a habitat on the island of Sumatra.

Ducks, geese, and other waterfowl feed on wetland invertebrates. A number of songbirds come to marshes and swamps to breed and to feast on insects. The overall abundance of plant and animal food attracts terrestrial mammals such as otter, raccoon, muskrat, and beaver. Few mammals, however, are restricted to wetland habitats.

Lake Biome. Lakes are relatively deep open bodies of water with little or no current. Small lakes are usually referred to as ponds, but no clear distinction between lakes and ponds can be made. Natural lakes are created through a variety of geomorphic processes. Most are of glacial origin and relatively young (see Plate XIII). Some of the world's largest and oldest lakes, however, were formed by tectonic processes, usually the rifting of the Earth's crust that resulted in deep troughs that later filled with water. Lake Baikal, the oldest, deepest, and largest (by volume) lake on Earth formed in this way. A few lakes are associated with the collapsed features of volcanoes known as calderas. Oregon's Crater Lake lies in one such caldera. Others result from cutoff bends or meanders along winding rivers. These crescent-shaped lakes are called oxbows—or, in Australia, billabongs.

Deep lakes often become stratified: their waters separate into distinct layers based on different temperatures and hence different densities. Oxygen and sunlight are available to organisms in the uppermost layer, but at depth both can be lacking. The lowest level, the hypolimnion, frequently exhibits such low oxygen levels in deep lakes that few organisms can survive there. Only mixing of the layers can carry oxygen, dissolved from the atmosphere, into the deep water. Shallow lakes are readily and frequently mixed by the wind. Deeper lakes may be mixed season-ally by winds or by the changes in water temperature that accompany changing seasons.

Several factors determine the types and abundance of life in lakes: water tem-perature, the depth to which light can penetrate, oxygen content, nutrient availabil-ity, and food sources. Generally three distinct life zones are recognized: the littoral, the benthic, and the pelagic.

The littoral or near-shore zone is the area in which rooted emergent plants such as cattails and reeds can grow (see Figure 1.14). Some limnologists extend the zone from the shoreline in deeper lakes to include those depths reached by sunlight suffi-cient to support the growth of submerged plants such as coontail, watermilfoil, and water weeds. Adequate light characterizes this shallow-water zone, as does abun-dant oxygen. Variations within the littoral zone depend on the nature of the sub-strate. The main distinctions are between rocky shores and soft-sediment shores, where marshes of emergents may develop. Soft-sediment littoral habitats provide safety for juvenile fishes and other small animals. The larvae of many insects also inhabit these areas. Adult waterboatmen, water sticks, and diving beetles are life-long residents. Fish are more abundant in this zone than in the other two, most feeding on the abundant invertebrates or on fishes smaller than themselves. Frogs and turtles are common. Larger predators, including many wading birds, are attracted to the cornucopia of prey species.

Figure 1.14 The littoral vegetation at the edge of a pond in Maine reflects the zonation of plants typical of lakes and wetlands. Farthest from shore are floating plants, such as pond lilies. Closer to shore are a ring of emergents such as pickerelweed. Lining the shore are shrubs and bog plants. *(Photo by author.)*

The benthic zone is the lake bed. Beyond the littoral zone in deeper waters, the lakebed is usually covered with fine sediments rich in organic matter. Benthic life consists mostly of microscopic decomposers (bacteria and fungi); filter-feeding detritivores such as midgefly larvae, nemotodes and other worms, and molluscs; and the animals that feed upon them, including fishes such as carp and catfish.

In the pelagic zone, water is too deep for rooted plants to survive, but a rich phytoplankton may develop if nutrients are sufficient. Cyanobacteria, diatoms, and dinoflagellates will be abundant. Zooplankters that feed on the phytoplankters will also be abundant and will be eaten by larger zooplankters, including tiny predatory crustaceans such as water fleas and copepods. Small fish eat zooplankters and become prey for larger fish. Larger fish feeding and swimming in the upper pelagic zone are caught by mergansers, grebes, loons, and Osprey and other fish-eating eagles.

A special category of lake is composed of salt lakes. Many are ephemeral, forming only after heavy rains and then drying up. None have river outlets, so water loss results only from evaporation. When the water evaporates, the salts dissolved in it precipitate out and over time accumulate in thick deposits of the floor of the basin. Some salt lakes were large freshwater bodies of water during cooler, moister episodes during the Pleistocene Epoch, but slowly evaporated as the climate warmed and became drier. Great Salt Lake, Utah, is the remnant of such a lake that in recent decades has shrunk and expanded in response to regional climatic cycles.

As a lake's water volume fluctuates so does its salinity and usually its oxygen levels. Organisms living in these environments must tolerate changing salinity as well as dessication when the lake dries up, and submergence when the lake bed fills. Generally, the higher the salinity the fewer species inhabit the lake. Bulrushes, saltgrasses, and alkali grasses occur regularly in the littoral zone of saline lakes. Submerged plants are limited to such plants as sago pondweed, wigeongrass, and spiral ditchgrass. Some of these are the same species found in saltmarshes and sea-grass meadows in marine biomes.

Brine shrimp are one of the few invertebrates to thrive in saline waters, and the populations may be huge. Some midges, damselflies, and dragonflies also tolerate high-salinity waters. Most fish cannot tolerate such extreme conditions, the exception being some lungfish that burrow into the mud and slow down their metabolism until fresher water returns with the rains. Invertebrates and algae may encyst, encapsulating themselves in a waterproof cover, and go into a resting or dormant state.

Although relatively few species live in saline lakes, the populations of those that do may be enormous at certain times of year. The abundance of brine shrimp and insect larvae attract wading birds and waterfowl, especially during migration. Perhaps most spectacular are the millions of flamingoes that flock to breed in the shallow salt lakes of East Africa.

The Marine Biomes

Coastal Biome. The edges of the land, coasts, are affected by saltwater every day. The landward boundary of a coast is marked by a zone in which terrestrial plants and animals are repeatedly showered by salt spray from breaking waves. The seaward boundary lies offshore at a depth of about 200 ft (60 m), the depth at which wave action rarely if ever disturbs the seabed. Organisms that thrive in coastal habitats must be adapted to a complex set of variables, most important wave action, varying times of exposure to the air and submergence below saltwater, and conditions imposed by the substrate, whether it is solid rock or fine sediments or any particle size in between. Latitude, since it affects climate, is another significant factor. Tropical coasts provide different conditions for life than do mid-latitude or polar coasts. Taken altogether, coasts may contain the largest variety of habitats and microhabitats of any of the world's biomes (see Figure 1.15 and Plate XIV). And a host of organisms—microorganisms, plants, and animals—successfully occupy each.

All coasts are characterized by life zones arranged at different heights above mean sea level or depths below mean sea level. Each zone has its own typical assemblage of organisms, although different substrates and geographic areas will have different species present. These life zones are quite obvious on rocky coasts, but difficult to see on sandy shores. The uppermost zone, the spray zone, is properly called the supralittoral zone. (Littoral means shore. The same term is used to designate the shore zone of a lake.) In this zone, salt spray is a major factor. The supralittoral is wetted by waves only during major storms; its lower limit occurs at the high-tide mark. The species-poor community is largely made up terrestrial

Figure 1.15 Coasts such as here in the Galapagos Islands consist of many different habitats, from sandy beaches to rocky headlands. Here in the tropics, the intertidal zone may be occupied by mangroves, and the shallow waters just offshore may host coral reefs. *(Photo by author.)*

organisms. Few land plants tolerate salty mists and soils. On rocky coasts, the lower part of the zone—sometimes referred to as the supralittoral fringe—is covered with a distinct band of black lichens and a coating of cyanobacteria. Some perennial seaweeds, true marine species, may also grow in this zone. The characteristic animals of the lichen-cyanobacteria belt are periwinkles (grazers) and isopods (detritivores). Grapsid crabs and hermit crabs, insects, and small birds visit these areas to feed. Steep headland cliffs provide safety from most larger predators and often become favored nesting sites for seabirds. Thousands of fulmars, puffins, murres, or gannets may crowd together to lay their eggs on the narrow ledges; but these birds feed out at sea and not in coastal habitats.

The intertidal zone or eulittoral zone encompasses coastal areas between the highest high tide (spring high tide) and lowest low tide (spring low tide). It usually consists of three distinct bands: the upper-shore, mid-shore, and low-shore zones. On rocky coasts in the mid-latitudes the upper shore is the barnacle zone, although it also hosts brown and red algae and a biofilm of cyanobacteria. Mussels are common on extremely exposed shores. Both of these invertebrates are sessile or attached filter-feeders. The most common motile animals are limpets, grazers of encrusted red algae and cyanobacteria. The main predators are various whelks. The mid-shore and low-shore zones are home to green, red, and brown algae and beds of mussels. The mussels themselves create habitat for animals that attach to them or bore into their shells, and they may host algae, barnacles, anemones, and hydroids. Clumps of mussels trap sediments and create a microhabitat suitable for polychaete worms. Limpets and chitons move into the mussel colony to graze algae. Various isopods, amphipods, and shrimps clean up organic detritus. And predators such as sea stars, crabs, fishes, and seabirds pry invertebrates from the rocks.

The subtidal or sublittoral zone is submerged beneath seawater most of the time and exposed only during spring low tides. In cooler waters this zone is usually occupied by kelps, large brown algae of the order Laminariales. In warmer waters, tunicates and red algae coat the rocks.

Soft-sediment coasts are three-dimensional habitats; along with the horizontal zones so apparent on rocky coasts, a vertical zonation of lifeforms develops along these coasts. Some organisms live *on* the beach but many more live *in* the beach. All must contend with an unstable substrate. The swash and backwash of waves constantly moves small particles. The animals of the beach community move articles around as they dig, burrow, and feed. Attached forms of both algae and invertebrates, so common on rocky coasts, are essentially absent on sandy ones. They cannot gain a steady foothold. Furthermore, organisms have to be highly mobile to respond rapidly to changing conditions of exposure and submergence associated with the tides. Photosynthesis is done almost entirely by cyanobacteria, diatoms, and flagellates. Many live within the spaces between sand grains and undergo vertical migrations, coming to the well-lit surface during daytime low tides and sinking back into the sand when sunlight is not available during high tides and at night. Fine muds and sands are not easily flushed of dead algal cells and wastes;

they become rich in organic content and support bacterial and fungal decomposers. Very small invertebrates such as rotifers, ostracods, nematodes, and some cope-pods graze on the bacteria or consume detritus. The larger invertebrates on sandy beaches include bristle worms (polychaetes), both filter-feeders and deposit-feeders; crustaceans such as isopods, amphipods, crabs, and ghost shrimp; echinoderms; and molluscs, including deposit-feeding gastropods, carnivorous nudibranchs, and a variety of bivalve clams.

The abundance of invertebrates attracts predators. Shorebirds such as sand-pipers, plovers, and oystercatchers continually probe the sands to find buried mor-sels of food and to discover what the sea may have washed in to the shore. The occasional stranded sea mammal is a bonanza for scavengers of all sorts.

A number of other major habitats exist on coasts and serve as vital interfaces between the land and the sea. Estuaries, mudflats, salt marshes, and mangroves all protect the land from erosion and act as sheltered nursery areas for many marine invertebrates and fishes. In each habitat, the zonation generally so characteristic of the Coast Biome is evident. Generally the zones replace one another along salinity and moisture gradients running from the land to the sea.

Continental Shelf Biome. The continental shelf is the permanently submerged outer portion of a landmass. It is an extension of continental bedrock and not, geo-logically speaking, true sea floor. A shelf begins at the extreme low-tide mark and extends out to sea to a depth of about 600 ft (200 m). Many shelf areas are quite narrow, but the widest may extend 900 mi (1,500 km) offshore. The seabed and the relatively shallow waters above the shelf are some of the most productive and eco-nomically important areas of the ocean. As much as 90 percent of the world's catch of shellfish and finfish come from this biome. The richness of shelf waters also sup-ports many seabirds and marine mammals.

Continental shelf communities depend on nutrients washed in from the land and carried in from the open sea by tides and currents. The seasonal pulse of river runoff is full of nutrients and stimulates seasonal algal blooms in shelf waters. Since the shelf waters are shallow, however, they usually become stratified in summer in the mid-latitudes and all year in the tropics. Warmer surface water floats on top of the cooler, denser water lower in the water column. Particles, be they phyto-plankters or particulate organic matter (POM), tend to sink out of the sunlight sur-face waters and settle to the seabed. Mixing of the water column is necessary to keep nutrients in the sunlit layer where the phytoplankton can absorb them. In the waters above the continental shelf, mixing is achieved by wind, tides, fronts, and upwelling. Wherever nutrients are concentrated, primary productivity is high. Most zooplankters are copepods. They themselves are food for krill and small fish, which in turn are consumed by carnivorous fish, seabirds, and whales. Actual com-munity members vary with latitude and water temperature. Five separate shelf environments are significant in terms of high primary productivity: seagrass mead-ows, kelp forests, banks, regions of upwelling, and coral reefs (see Plate XIV).

The continental shelf merges with the subtidal or sublittoral zone of the Coast Biome on its landward edge. Where the seabed is sandy, tube-dwelling and burrowing polychaetes invertebrates are common. Organisms such as molluscs, sea cucumbers, sea urchins, crabs, flatfish, and stingrays abound. In sheltered locations at the heads of estuaries in lagoons, on the lee sides of barrier islands, seagrasses may grow in dense underwater meadows. These are true flowering plants rooted in the substrate, so they require shallow water where sunlight is readily available. The seagrasses may bind fine particles and stabilize the sands, permitting filter-feeding oysters, mussels, and clams to gain a foothold. The fecal pellets of these molluscs are consumed by deposit-feeders. Demersal fish, those that live and feed near the bottom, are plentiful.

Few animals consume seagrasses directly, but many invertebrates graze the epiphytic diatoms and filamentous algae that cover their blades. The main food chains in seagrass meadows are detritus based. In the substrate polychaete worms are dominant; on the surface crabs, shrimps, amphipods, and fish are the chief detritivores. An abundance of shellfish and fish draws in shorebirds, diving ducks, fish-eating eagles, and Ospreys.

Banks are underwater plateaus rising to shallow depths on the continental shelf. They obstruct the flow of ocean currents creating upwelling and hence nutrient-rich waters. Some of the great fisheries of the recent past are associated with banks where demersal fish were the most valuable catch. The largest and best known are the Grand Banks off Newfoundland, Canada. Until the fisheries collapsed, these banks supplied Atlantic cod, haddock, ocean perch, turbot, flounders, and plaice, as well as crabs, shrimps, and scallops. The once enormous cod and herring populations supported some 30 different marine mammals.

Off rocky coasts where waters are cool, kelp forests develop. If the slope of the continental shelf is gradual, these forests extend as much as 6 mi (10 km) offshore. Kelps are large brown algae, some of which grow 200 ft (60 m) long; they create a three-dimensional habitat resembling a true forest on land. Kelp forests are highly productive ecosystems. Pieces of kelp eroded from the main plant by waves start a detritus food chain as bacteria decompose the fragments. The bacteria cycle nutrients through a microbial loop but are also food for zooplankters and small filter-feeders. Unconsumed bacteria and algae settle to the bottom where they support a rich benthic community of sessile invertebrates, including mussels, barnacles, sponges, and tunicates. Sea urchins are consumers of live kelp, although most of the time their impacts are negligible. On occasion, however, urchin populations explode and devastate the kelp beds. Sometimes this happens when their predators, such as sea otters, are in low numbers, but not always.

Cool waters are also associated with regions of upwelling. Where winds blow parallel to and away from west coasts in the subtropics, the warm surface waters are driven off and cold water from depth rises to replace it. The upward flowing water carries nutrients back to the surface. The result is a cold ocean current and productive marine ecosystem. These conditions exist with the Humboldt Current off western South America, the Benguela Current off southwest Africa, the California Current off California, and the Canary Current of northwest Africa. A similar situation develops seasonally in the northwest Indian Ocean in response to the Asian monsoons. In each instance a pelagic fishery, especially of anchovies, has supported both subsistence and commercial interests. The climate associated with cold currents and their regions of upwelling is typically arid. Indeed some of the world's driest deserts occur adjacent to these oceanic features. This too has proved of great economic importance. The thousands if not millions of seabirds that gather on offshore islands to gorge on anchovies deposit their droppings, that is, guano, on the headlands and islands. With no rain to wash it away, the guano accumulates to considerable thickness. Before the development of synthetic fertilizers, phosphorus-rich guano was in great demand by farmers in Europe and North America and most of it was mined from offshore islands. Now less is being produced. The fisheries have collapsed, apparently due to a combination of overfishing and climate change, and the guano bird populations—cormorants, pelicans, and boobies—have crashed. In the Southern Hemisphere, penguins used to burrow into the guano deposits to nest. Their populations are also declining due to fewer fish and fewer nesting sites.

Coral reefs are famous for the enormous variety of invertebrates and fishes that inhabit or visit them. More than 100,000 species have been recorded to date. Their great species richness has led some to equate them with the tropical rainforest on land. Coral reefs are great limestone structures built by stony coral polyps over thousands of years. They develop only in warm, clear (low-nutrient), shallow tropical waters. The most diverse reefs are located in the Indo-West Pacific biogeographic region.

Just as rainforests grow on nutrient-poor soils, coral reefs form in low-nutrient marine "deserts." The same trick of tightly recycling nutrients is evident in both, conserving and concentrating whatever is available to the reef community. The phytoplankton plays a small role in primary production. Most solar energy is fixed by algal turfs, coralline algae, or zooanthellae, the dinoflagellates embedded in coral tissue and existing in a symbiotic relationship with the polyps.

In addition to the variety of forms and colors of the corals themselves, what most often attracts tourists and divers to reefs are the fish. More than 4,000 species from 100 different families are associated with coral reefs. Many are brightly colored and bear distinctive patterns. Most widespread are damselfish, parrotfish, surgeonfish (or tang), wrasses, Moorish idols, butterfly fishes, and angel fishes, all fishes confined to reef habitats. They feed variously on phytoplankters and zooplankters or graze the algal turf; a few prey on coral polyps. The huge numbers of small fish attract larger seagoing predators such as groupers, snappers, jacks, barracudas, and sharks and rays. At the reef, these larger animals also seek the cleaner wrasses, gobies, and shrimps that remove their external parasites.

A final group of organisms associated with coral reefs are the bioeroders. These animals break down the reef, reducing it to coral sands and rubble. These animals either bore into the reef foundation or gnaw away at the coral and coralline algae. Among the borers are sponges, polycheate worms, and bivalves. Working away at the surface are chitons, urchins, limpets, hermit crabs, pufferfish, and parrotfish.

The delicate habitats of coral reefs are threatened by warming ocean temperatures, increased amounts of nutrient and sediments washing in from nearby shores, destructive fishing practices, physical damage from tourists, and the jewelry trade.

Deep Sea Biome. Beyond the continental shelf lies the deep sea, a vast region covering 65 percent of the Earth's surface. Water depth ranges from 650 ft (200 m) below sea level at the edge of the shelf to a maximum of 36,198 ft (11,033 m) below sea level at the bottom of the Mariana Trench. Most of the sea floor is a vast abyssal plain covered with muds and calcareous or siliceous oozes. Rising above the plain are mid-oceanic ridges and seamounts, hard surfaces providing habitat for sessile creatures. This is the least-known part of the Earth. Exploration continues and information improves with every advance in underwater vehicular and sampling technologies.

The deep sea is an extreme environment, but one that is fairly stable. Cold, darkness, and high pressure require special adaptations. Salinity is almost always a constant 35. Oxygen is available in adequate amounts throughout the water column. Oxygen dissolved by surface waters is carried to depth on a massive conveyor belt and distributed by slow deep sea currents.

The darkness of deep waters precludes photosynthesis, so living phyoplankters are absent. Yet their production in well-lit surface waters is integral to most food chains on the sea floor, which are necessarily detritus based. POM and DOM, as well as carcasses of fish and marine mammals, are major sources of energy and nutrients. Common members of the community are foraminiferans, ncmatodes, copepods, polychaetes, bivalves, and isopods, as well as larger forms such as sea anemones, brittlestars, sea stars, sea cucumbers, and fishes. Most of these are deposit-feeders, but suspension-feeders such as glass sponges, horny corals, and sea pens are also present. Where hard surfaces are available, deep sea coral communities may develop. Only 10 years ago, the existence of these animals was unknown, but it now seems that more types of coral may live in cold deep seas than in warm tropical seas.

Some of the communities of deep sea organisms just now beginning to be studied include those associated with seamounts, hydrothermal vents, and cold seeps. Seamounts are underwater mountains rising high into the nutrient-poor waters of the open sea. They promote localized upwelling and their shallow summits concentrate nutrients and phytoplankters above them. Zooplankters thrive and attract predators such as mysid shrimp and squid. Open sea predators such as sharks, rays, orange roughy, tuna, and swordfish come in to feed on the small invertebrates. Deeper seamounts support more suspension-feeding organisms, including stony corals, horny corals, black corals, sea anemones, sea pens, hydroids, sponges, tunicates, and crinoids. Coral forests develop on rocky outcrops where fast currents remove wastes and carry in food particles. The coral forest offers protection and sites above the sea floor for other suspension-feeders, so a rich community of invertebrates and their vertebrate predators forms.

Hydrothermal vents and cold seeps have become famous as ecosystems where the primary energy comes from chemicals not sunlight. Bacteria are the chemosynthesizers, extracting energy from sulfur compounds released at the vents. Some are free-living, but many have evolved symbiotic relationships with invertebrate hosts: vestimentiferan tubeworms, vesicomyd clams, and bathymodilid mussels. (Those long "first" names refer to the way taxonomists have classified them.) Tubeworms are completely dependent on bacteria housed in special tissues. The clams nurture bacteria in modified gills but are also filter-feeders despite greatly reduced digestive systems. The mussels have fully functional digestive tracts, but the food they filter from the sea seems only to supplement that produced by the bacteria in their gill tissues. Some shrimps also have symbiotic relationships with sulfur-dependent bacteria. At cold seeps, the energy source is methane. Mussels and three other families of bivalves are currently known to have symbiotic relationships with methane-using bacteria.

Predators at seeps and vents include eel-like zoarchid fish, limpets, and amphipods. Crabs and squat lobsters target mussels and tubeworms at vents, and octopuses prey on clams, mussels, and crabs.

A final deep sea habitat that is receiving attention is one created by whale and other large animal carcasses and skeletons that have settled to the sea floor. These are random nutrient-rich flecks on a vast sea floor. Scavengers such as demersal fish, amphipods, decapod shrimps, gastropods, and brittlestars colonize them relatively quickly and may remain until the flesh has been consumed. Carnivores arrive to feed on the scavengers. Skeletons host chemosynthetic communities of clams and mussels. White and yellow filamentous bacterial mats eventually may cover the bones and support populations of limpets. Amphipods, isopods, and polychaetes are also known to be members of carcass communities. It has been hypothesized that randomly placed carcasses and skeletons on the ocean floor may serve as stepping stones for the migration of organisms from one cluster of hydrothermal vents to another when these short-lived phenomena of mid-oceanic ridges go extinct and new ones appear somewhere else.

The deep sea, of course, has a water component as well as the bottom habitats mentioned above. The open ocean is divided into a number of depth zones in which pressure and temperature impose increasing degrees of challenge to life. Biodiversity is relatively low. The epipelagic zone includes depths from the sea's surface down to about 600–850 ft (200–250 m) below sea level. Sunlight penetrates this zone, but the phytoplankton is limited by the low nutrient levels found so far from the world's landmasses. Most zooplankters rise to the surface at night to feed when it is safer, since their predators, even copepods, hunt by sight. Many zooplankters are transparent and nearly invisible to the nocturnal hunters of the sea.

The next zone is the mesopelagic, extending down to 3,200 ft (1,000 m) below sea level. No living phytoplankters exist here, so animals are either detritivores or carnivores. Copepods and gelatinous siphonphores are abundant. Shrimps and other animals are frequently red or orange. At the surface, these colors would make them highly visible, but at depth, it renders them invisible. Red and orange pigments absorb the only wavelengths of light that penetrate this far, so the animals essentially become "black." Many fishes bioluminesce: they produce light chemically in specialized organs. Their flanks are also highly reflective. Both features disrupt the outline of their bodies, making them more difficult to see by predators swimming beneath them and viewing them against the bright disk of sunlight (Snell's circle) at the surface, still visible at these depths.

In the bathypelagic zone between 3,200 and 8,000 ft (1,000 and 3,000 m), fish typically are black over their entire bodies. Their skeletons are poorly developed and they lack air-filled swim bladders. Muscles are also greatly reduced and most predators lie in ambush rather than waste energy chasing after prey. Fish have small eyes or are blind but have very large mouths. Most produce bioluminescent compounds. In this dark habitat, it may be that light aids species recognition, much as color and patterning do in well-lighted parts of the ocean. Some fish use light as a lure to attract prey; others use it to create decoy targets and divert the attack of a predator.

In the abyssopelagic zone below the bathypelagic level, food is limited and life sparse. Few fish inhabit the region. Decapod crustaceans are generally dominant, and in the deepest parts, mysid shrimp are particularly abundant.

The last zone in deep sea waters lies within 300 ft (100 m) of the sea floor and is named the benthopelagic zone. Food is more abundant in this zone than in the layer immediately above because bottom-dwelling larvae of gastropods, amphipods, and sea cucumbers float up off the seabed and may be consumed by swimming organisms.

Biomes versus Ecoregions

Biomes have been criticized as being too vaguely defined and determined, leaving sizes and boundaries open to interpretation. In part this weakness in the concept is due to the use of vegetation to determine the limits of a particular biome. On maps, the separation between neighboring biomes is a sharp line; in reality, a zone of transition from one vegetation type to another is the rule on the ground. One attempt to overcome this perceived problem is becoming increasingly popular among people charged with managing and protecting natural areas and that is the concept of an ecoregion, developed by Robert Bailey of the U.S. Department of Agriculture, Forest Service.

"Ecoregion" is a contraction of "ecosystem region." The emphasis is on region or area rather than on growthforms or communities of plants and animals. The idea is that geographic areas should be defined first according to natural features, particularly those that control or differentiate ecosystems. Significant change in these controls tells the mapmaker where boundaries should be placed. This is promoted as a more objective approach than mapping vegetation and one that better approximates nature. The major control of ecosystems is climate, itself influenced by latitude, position on a continent, and altitude. Climate determines how the hydrologic cycle works in a particular region and therefore is of prime importance in controlling moisture regimes that are so significant in the annual rhythms of life. Climate and moisture are major players in landform development, erosion cycles, and soil formation processes. They influence fire frequency. And they are key elements in potential primary productivity rates. All of these factors work to determine the vegetation and the structure and functioning of ecosystems. Ultimately, vegetation becomes a key indicator of climate, as it is the case in the biome concept.

Ecoregions are conceived as existing at various scales, and this is the real contribution of the concept to practical applications. At the macroscale, an ecoregion covers about 40,000 mi^2 (100,000 km^2). This creates a somewhat finer-grained division of Earth's major ecological systems than the biome, but it is roughly the same thing. Bailey's method revealed 15 such terrestrial ecoregion *divisions*, as these macroscale units are called, compared with the eight or nine commonly used in biome schemes (see Table 1.2).

Major ecoregions of the oceans are termed domains. The three domains are Polar, Temperate, and Tropical. The physical characteristics of ocean waters control

Table 1.2 Comparison of Major Biomes and Ecoregions

Biome	Ecoregion
Tundra	Polar Domain
	Tundra Division
Boreal Forest (taiga)	Subarctic Division
Temperate Broadleaf Deciduous	Humid Temperate Domain
Forest (including mixed forests)	Warm Continental Division
	Hot Continental Division
	Subtropical Division
	Marine Division
Mediterranean Woodland and Scrub	Mediterranean Division
Temperate Grasslands	Prairie Division
	Dry Domain
	Temperate Steppe Division
Desert	
Cold Deserts	Temperate Desert Division
Hot Deserts	Tropical/Subtropical Desert Division
Tropical Savanna	Tropical/Subtropical Steppe Division
	Humid Tropical Domain
	Savanna Division
Tropical Seasonal Forests	Savanna Division
Tropical Rainforest	Rainforest Division

Table 1.3 Oceanic Ecoregions

Polar Domain
  Inner Polar Division
  Outer Polar Division
Temperate Domain
  Poleward Westerlies Division
  Equatorward Westerlies Division
  Subtropical Division
  High Salinity Subtropical Division
  Jet Stream Division
  Poleward Monsoon Division
Tropical Domain
  Tropical Monsoon Division
  High Salinity Monsoon Division
  Poleward Trades Division
  Trade Winds Division
  Equatorward Trades Division
  Equatorial Countercurrent Division
Shelf

the distribution of ecoregions in the sea. Major differences in water masses involve temperature, salinity, and color. Each domain is subdivided into two or more divisions, each largely defined according to the global wind system that moves ocean currents and distributes heat, nutrients, and organisms around the world ocean. The end result (see Table 1.3) is a system resembling that proposed by Alan Longhurst in 1998 as major marine biomes, with more categories. Whereas Longhurst based his biomes on temporal patterns of life, namely phytoplankton blooms, Bailey bases his divisions on wind and current patterns, which are major controls of the distribution of living organisms and processes.

At the mesoscale, ecoregions are based on landform distinctions and are called landscapes. Each corresponds to an area of about 400 mi^2 (1,000 km^2). Microscale units are *sites*, each about 4 mi^2 (10 km^2) in size. These are delineated according to edaphic criteria, including slope, aspect, ground conditions, and local geology. The two smaller scales are well suited to their intended purpose of allowing assessment of different resources on the same unit of land. By classifying and mapping land units rather than resource distribution areas, the piece of land under consideration is always the same. If boundaries were drawn along the limits of each resource, any number of unique but overlapping areas would be the result, complicating planning and monitoring multipurpose use of public lands and other natural areas. The system has been adopted by the U.S. Forest Service, U.S. Geological Survey, and the World Wildlife Fund, among others.

Further Readings

Books

Overviews of terrestrial biomes are common in ecology and physical geography textbooks. Detailed coverage of both terrestrial and marine biomes may be found among the other titles in the Greenwood Guides to Biomes of the World.

Bailey, Robert G. 1996. *Ecosystem Geography.* New York: Springer-Verlag. (Presents Bailey's system of ecoregions.)

Internet Sources

The following Web sites are informative and well-illustrated sites on biomes:

Benders-Hyde, Elisabeth, and Karl Nelson. 2000–2002. "Blue Planet's World Biomes." http://www.blueplanetbiomes.org/world_biomes.htm. This site covers terrestrial biomes and was prepared by students and teachers at the West Tisbury School, a kindergarten through eighth grade school on Martha's Vineyard, Massachusetts.

Missouri Botanical Garden. 2005. "What's It Like Where You Live?" http://www.mbgnet.net. Covers terrestrial biomes and freshwater and marine ecosystems.

University of California Museum of Paleontology. 1994–2008. "The World's Biomes." http://www.ucmp.berkeley.edu/exhibits/biomes/index.php. Covers both terrestrial and aquatic biomes.

Woodward, Susan L. 1996. "Introduction to Biomes." http://www.runet.edu/~swoodwar/CLASSES/GEOG235/biomes/intro.html. Created by the author of this book to support a course in biogeography at Radford University in Virginia. Covers only terrestrial biomes.

The following Web sites are organized according to ecoregions:

National Geographic and World Wildlife Fund. 2001. "WildWorld." http://www.nationalgeographic.com/wildworld/terrestrial.html. Maps and detailed information on the plants and animals of 867 land-based ecoregions, including mangrove. Emphasis is on threatened species. May also be accessed through the World Wildlife Fund Web site at http://www.worldwildlife.org/wildworld.

World Wildlife Fund. 2006. "Selection of Freshwater Ecoregions." http://www.panda.org/about_wwf/where_we_work/ecoregions/about/habitat_types/selecting_freshwater_ecoregions/index.cfm. Presents outstanding examples of freshwater ecoregions where the World Wildlife Fund is engaged in research and monitoring to help conserve global biodiversity.

World Wildlife Fund. 2006. "Selection of Marine Ecoregions." http://www.panda.org/about_wwf/where_we_work/ecoregions/about/habitat_types/selecting_marine_ecoregions/index.cfm. Presents outstanding examples of marine ecoregions where the World Wildlife Fund is engaged in research and monitoring to help conserve global biodiversity.

2

Patterns of Life

Biomes and their regional expressions are composed of plants and animals that have had long, independent histories of evolution and dispersal but currently inhabit the same geographic area. They may have originated in the same place or in different places, in the place where they are now found or on some distant landmass or in a faraway ocean. Ecologists talk of communities assembling, meaning that the component species come together to form an interacting group of organisms that function together to ensure the flow of energy and cycling of nutrients through the ecosystem. This process happens on local, regional, and global scales.

As species originate, move, are halted by or overcome barriers, adapt to new environments or go extinct, a variety of species distribution patterns result. Early naturalist-explorers, especially in the eighteenth and nineteenth centuries, saw the main outlines of these patterns. Modern scientists continue to search for their explanations. All of the patterns of life discussed in this chapter are important factors that make one biome distinct from another or one regional expression of the same biome distinct from all the others. The kinds and numbers of species or other taxa present, the evolutionary and geographic origins of the major plants and animals of a biome, the dominant growthforms of plants, characteristic mosaics or zonation of vegetation—all are major considerations in the study of biomes and each is discussed in this chapter.

Taxonomic Patterns

Taxonomy

Before the modern concept of a species developed, each kind of organism was recognized as conforming to a named specimen or "type." Those individuals deemed similar enough in appearance to be the same kind of plant or animal were classified as members of the same species. If two species were similar but not identical, they were placed together in a higher and broader category, the genus. Each species was seen as part of the higher category, a member of some genus. The official name of the species used both the genus name and the species name, both written in Latin. By convention, the genus name comes first and is capitalized; the specific name comes second and begins with a lower case letter. Both words are italicized, as appropriate for foreign language terms. The genus name may be used alone to refer to all its member species together, but the species name should never be used alone, since it has no real meaning outside the genus to which it belongs. The same specific name may be used in any number of different genera (the plural of genus). For example, *Cornus canadensis* is the bunchberry, a common forb in the herb layer of temperate deciduous broadleaf and mixed forests in North America; *Wilsonia canadensis* is the Canada Warbler, a Neotropical bird that breeds in the boreal forests of Canada; *Castor canadensis* is the beaver, and *Ovis canadensis* is a bighorn sheep. As in these examples, the species name indicates something about the geography of the bird or the region from which specimens were first collected and sent to taxonomists for classification. Species names also may indicate some characteristic of the plant or animal or honor some person involved in its discovery or simply respected by the taxonomist.

Linnaeus devised a system for categorizing and ranking organisms according to common physical traits that formed the basis for the taxonomic hierarchies used today. In his time, similar traits in different organisms were viewed as hints of the divine design of the universe. Later they came to be seen as indications of the common origins of groups of species. The concept of species changed from that of an unchanging type to that of a product of an evolutionary history or phylogeny. Two or more species placed in the same genus are considered descendents of the same ancestral species. Lines of descent extend back in time. Similar genera also descend from a common ancestor (the founder of a family), but one that lived farther back in evolutionary and geologic time. The idea of common descent allowed taxonomists to refine Linnaeus's ranking of different categories and gave them a more or less objective way of assigning a species to a nested set of taxa (levels in the taxonomic hierarchy) based on evolutionary relationships. All genera with a common ancestor are placed in the same family; all families with a common ancestor are placed in the same order, and so forth.

The more inclusive or generalized the group, the higher the taxon level (see Table 2.1). Several schemes currently exist and that affects the number and names of the highest levels. At the top of some schemes is the Domain, three of which are

● ●

Linnaeus and Taxonomy

Taxonomy is the science that classifies organisms. The modern classification system traces back to Carl von Linné (1707–1778) or Carolus Linneaus, the Latinized form of his name by which this Swedish botanist is best known. Linnaeus's interest in plants was an outgrowth of his medical studies, for eighteenth-century doctors had to know how to identify, prepare, and prescribe medicinal herbs. A love of nature combined with religious beliefs in natural theology led Linnaeus to try to reveal the divine order of God's creation by classifying all living things. The first edition of his *Systema Natura* was published in 1735, the same year he completed his medical degree.

Before Linnaeus, species names were wordy descriptions in Latin, and one naturalist's description did not necessarily agree with another's. Linnaeus simplified the naming (nomenclature) of organisms by preferring a binomial name for each species. His influence spread in part because of his students, many of whom he sent on trade and exploration missions all over the world. Daniel Solander, the naturalist on Cook's first circumnavigation of the world, brought the first Australian plants back to Europe (and to Linnaeus). Pehr Kalm collected plants in the American colonies. Other students traveled to South America, Asia, and Africa. As new information came to him, Linnaeus revised his thoughts and updated *Systema Naturae*. The tenth edition used the binomial system exclusively and became the international standard for nomenclature.

Plant and animal species named by Linnaeus himself still bear his name in formal literature, where the taxonomist is often acknowledged after a species' scientific name. The Bald Eagle, for example, is *Haliaeetus leucocephalus* Linneaus. After his death, his books, collections, and manuscripts became the foundation of the Linnaean Society of London, a prestigious organization that still encourages the study of natural history.

● ●

usually recognized: Eubacteria, Archaea, and Eukarya. Major subgroups within each Domain are Kingdoms, either five or seven are currently in use. Monera (sometimes simply called Bacteria) is the only Kingdom in the Domain Eubacteria and consists of single-celled organisms that lack a cell nucleus. Domain Archaea contains two Kingdoms, Crenarchaeota and Euryarchaeota. All multicelled species with cell nuclei are members of one of the four Kingdoms in Domain Eukarya: Protista, Plantae, Fungi, and Animalia.

Species in the various Kingdoms are divided into yet smaller groups or phyla (sometimes called divisions in plant taxonomy). Each of the approximately 38 phyla in Animalia is distinguished according to a basic body plan, such as the radial symmetry of echinoderms or the four-limbed pattern of chordates. Only 10 phyla of plants are recognized. Among them is the Phylum Bryophyta, considered the most primitive of plants since they lack the water-conducting (vascular) structures xylem and phloem. Bryophytes include mosses, liverworts, and hornworts. In all, the gametophyte (sexually reproducing) generation is more dominant in the life history than the sporophyte (asexually reproducing) generation that produces wind-dispersed spores. Plants once grouped together as Pteridophytes are now classified into four phyla: Psilophyta (whisk ferns), Lycopodiophyta (clubmosses and

Table 2.1 The Main Categories in the Taxonomic Hierarchy (from general to specific)

Domain
    Kingdom
        Phylum (or
        Division)
            Class
                Order
                    Family
                        Genus
                            Species
                              Sub-
                              species

quillworts), Equisetophyta (horsetails), and Polypodiophyta (true ferns). All are vascular plants that reproduce by spores. Another older grouping, that of "naked seed plants" still collectively called gymnosperms, has similarly been subdivided into several phyla: Cycadophyta (cycads), Ginkophyta (ginko), Gnetophyta, and Coniferophyta (cone-bearing vascular plants). Finally all the flowering plants or angiosperms are grouped together in the Phylum Magnoliophyta. This last group contains what are believed to be the most advanced and most recently evolved members of the Plant Kingdom and have seeds enclosed in ovaries.

Classes group together organisms with similar variations on the body plan or plant structural theme. For example, Phylum Chordata includes Class Aves (birds) and Class Mammalia (mammals) among others. Phlyum (or Division) Magnoliophyta is split into two Classes, Magnoliopsida (the dicots) and Liliopsida (the monocots). Members of each taxon have similar flower structures, leaf venation patterns, and seed structure.

Lower taxonomic groups have increasingly fewer characteristics in common and contain fewer and fewer species. The lowest level is the subspecies. These may be considered geographically distinct populations within a species and are identified by a trinomial that consists of genus, species, and subspecies names. In the study of biomes, the taxa of most importance are species, genus, family, and order. These groupings are usually enough to tell how similar or how different the organisms in separate biomes or their

Recognizing and Naming Species

The biological definition of a species is a group of similar individuals that breed only with each other and that produce offspring of the same kind as the parents. The species is a natural unit in that members of a species "recognize" each other and may have evolved specialized structures, colors, songs, or behaviors to ensure that they do not mate with members of other species. Every higher unit or taxon (the singular of taxa) in the taxonomic hierarchy is artificial: it has been determined by people, by scientists. International committees provide testing grounds for the soundness of classifications and names applied and only after approval has been given does the Latin or scientific name come into use. The advantage is that the name is now in universal use. Each binomial (two-part) name is attached to only one species so everyone can know exactly what it refers to.

regional expressions are. In the Tundra Biome, for instance, many species are the same all around the Arctic Ocean on both North America and Eurasia. Some of the same plants even extend southward along the Appalachian Mountain chain. In the Temperate Broadleaf Deciduous Forest Biome, the same genera of trees as well as some birds and mammals occur in Europe, China, and eastern North America, but different species inhabit each. English people thus experienced a general sense of familiarity with the forests of their 13 American colonies, enough to provide European common names to many plants and animals. Only when something was entirely new to them—such as raccoons and opossums—did they adopt the names used by Native American peoples. In the Tropical Rainforest Biome, not only are the species encountered different from those in temperate forests, but so are genera and even families. The tropics therefore seem like exotic places to North Americans and Europeans. Different families—in some cases different orders—occur in the different regional expressions of the Tropical Rainforest Biome. Most plants and animals found in South America, for example, are absent from Africa, and vice versa.

The geographic pattern of similarity or dissimilarity outlined above was interpreted by Buffon (Georges-Louis Leclerc, Comte de Buffon [1707–1788]), one of the early developers of the biome concept, to mean that the center of origin for most species lay in the far north and that most species originated during a period when the Earth was warmer than today. Animals moved south as the climate cooled and "degenerated" (today, they would be said to have evolved) the farther they moved from the center of origin. Thus, species (mammals were Buffon's main interest) on different continents were quite similar in the Northern Hemisphere (North America and Eurasia), but those in the New World Tropics were quite different from those farther north and had moved and changed along a separate migration route.

Today genetic indicators are used to determine evolutionary relationships and help taxonomists correctly categorize species. Some changes at high taxonomic levels have ensued, but most work actually substantiates the results of early scientists who worked with pressed flower parts, or study skins, or bones. Even in the past, the information from specimens was constantly reevaluated and sometimes relationships and hence the taxa into which a species was placed changed. As a consequence, scientific names in older literature may not be identical to those of contemporary articles and books. In taxonomic works, reference may be made to both names, which will be designated synonyms. With time, everyone will learn and accept the new names, which will be used until newer information requires a revision of thinking and naming.

Taxonomic Regions: Zoogeographic Provinces and Floristic Kingdoms
As new species were being discovered and cataloged in the nineteenth century, it became apparent that groups of families were limited to one part of the world or another, while other families were more widespread and shared among two or

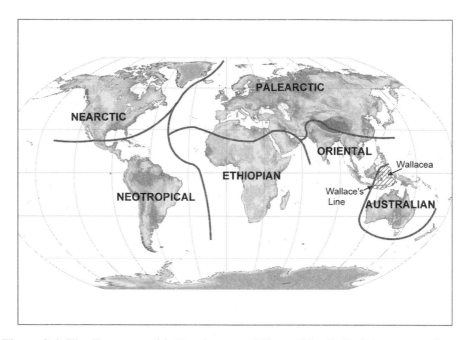

Figure 2.1 The Zoogeographic Provinces as delineated by P. L. Sclater according to the distribution of bird families. Later, Alfred Russel Wallace added what has become known as Wallace's Line dividing the Oriental and Australian Provinces. Today, it is generally recognized that a transition zone separates the two provinces, and the zone is named Wallacea in honor of the father of animal geography and co-founder of the theory of evolution by natural selection. *(Map by Bernd Kuennecke.)*

more regions. These patterns were seen as a way to classify geographic areas according to the taxa found in each area, irrespective of climate, soils, vegetation, or any other environmental condition. In 1857, the British ornithologist Philip L. Sclater (1829–1913) delineated six Zoogeographic Provinces (see Figure 2.1) based on the distribution patterns of bird families. He called them the Palearctic, Nearctic, Neotropical, Ethiopian, Oriental, and Australian Zoogeographic Provinces.

The boundaries between them corresponded to geographic barriers to bird movement that allowed a particular region to have a number of unique or endemic families as well as some that were shared with neighboring provinces. Other families in adjacent provinces were absent. A couple of decades later, Alfred Russel Wallace (1823–1913) worked out similar patterns for mammals and other animals and defined a clear boundary between the Oriental and Australian provinces, which today bears his name: Wallace's Line. Naturalists debated the boundaries and shifted them slightly from time to time; indeed, discussions still come up today. Many prefer to honor Wallace with a broad transition zone composed of the islands between the Malay Peninsula and New Guinea they call Wallacea (see Figure 2.1). Few if any of the boundaries are actually the sharp lines that appear on maps. The system works best for warm-blooded animals, that is, birds and mammals. Cold-blooded reptiles, amphibians, and fishes do not seem to play by the rules as well. The usefulness of the scheme lies in the names given to the large regions, each originally viewed as a center of origin for many taxa. The names provide handy terms by which to refer to various regions of the world and the species, genera, and families characteristic of them. Perhaps most commonly used is the term Neotropical, referring to southern Mexico, Central America, and South America and the animals most commonly associated with those areas. For example, in the United States a major conservation concern revolves around the decline of Neotropical migrants. These are birds that winter south of the U.S. border with Mexico, but come north to nest in the spring and summer. Neotropical songbirds such as thrushes, tanagers, and wood warblers are important functional and esthetic components of many North American (Nearctic) ecosystems.

The distribution patterns of plant families do not exactly mirror the Zoogeographic Provinces as devised for birds. Adolf Engler (1844–1930), a prominent German botanist of the late-nineteenth century, recognized four large regions he called Plant Realms. The Arcto-Tertiary Realm extended across the subarctic and temperate regions of the Northern Hemisphere. The Paleotropical Realm corresponded to the Old World tropics and stretched from Africa to northern Australia. The Neotropical Realm encompassed most of Central America and South America. The fourth realm, which Engler called the Ancient Oceans Realm, contained widely separated parts of the temperate and subantarctic zones of the Southern Hemisphere, including coastal Chile and Tierra del Fuego in South America; the Cape Region and south coast of South Africa; most of Australia; Tasmania; South Island, New Zealand; and the islands of the Southern Ocean. Ludwig Diels (1874–1945), another German botanist, subdivided the Ancient Oceans Realm into an Antarctic Realm (southern South America and islands in the Southern Ocean), a Cape Realm (surrounding the Cape of Good Hope, South Africa), and an Australian Realm. He removed New Zealand to the Paleotropical Realm. The system was reworked again in the middle of the twentieth century by the British botanist Ronald Good (1896–1992), who called the realms "Floristic Kingdoms" (see Figure 2.2). Good took Diel's Paleotropical and divided into three parts: Africa, Indo-Malayan, and Polynesian subkingdoms. He returned New Zealand to the Antarctic Kingdom, because it shares a flora of ancient conifers (podocarps and araucarias) and southern beeches (*Nothofagus*) with South America and Australia, landmasses once joined together as Gondwana. The most recent modifications have been by the Russian-Armenian botanist, Armen Takhtajan (1910–), a major figure in twentieth-century plant geography.

The tiny Cape Kingdom is sometimes challenged as a distinct Floristic Kingdom, both because of its size and the fact that, although diverse in terms of endemic species (73 percent of its flora of 8,850 species), only seven or eight plant families are endemic. Nonetheless, it has become conventional to talk of a Cape flora, so the descriptive value of a Cape Kingdom continues, as illustrated in

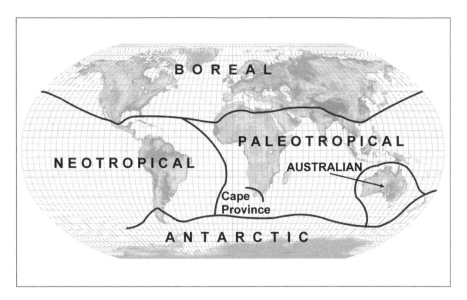

Figure 2.2 Floristic Kingdoms as set up by Ronald Good. The greater age of plants and better dispersal capabilities compared with animals makes these taxonomic regions somewhat different from the Zoogeographic Provinces. *(Map by Bernd Kuennecke.)*

discussions of the species-rich fynbos, the South African regional expression of the Mediterranean Woodland and Scrub Biome.

Patterns of Species Richness

Johann Reinhold Forster (1729–1798) was the naturalist aboard James Cook's second expedition to the South Pacific and as such was able to observe changes in plant life from the temperate regions of the Northern Hemisphere through the tropics and across the Equator into the higher latitudes of the Southern Hemisphere. He remarked on the steady decrease in the number of species with distance from the Equator. This spatial pattern is now well recognized and referred to as the latitudinal gradient in species diversity or, more simply, the diversity gradient. Forster believed species numbers increased or decreased with the latitudinal changes in surface heating. Why so many species live in the tropics and how they manage to coexist are still questions being investigated. One school of thought thinks that the tropics have been more or less stable through recent geological time, and stability allowed for a greater accumulation of species (assuming that climate change increases extinction rates). In contrast, lands in the middle and high latitudes were covered with ice several times during the Pleistocene Epoch, and their faunas and floras were decimated and have not yet had time to recover. Another school of thought considers the time factor in terms of ecological consequences. Species in the tropics have been able to evolve specialized niches that allow more to "pack" into a given ecosystem, whereas younger species in the middle and high latitudes have broader niches, so fewer can coexist. Niche is a functional concept relating to how a species obtains the resources it needs to survive long term. It was once equated with a person's job, how someone supports him- or herself. One person may be a generalist, a jack-of-all-trades, who lives by doing a bit of this and bit of that, not particularly expert in anything but good enough to do most of the needed work in a household or community. This person has a "broad niche." Another person may be a specialist doing one thing very, very well and have a "narrow niche." He or she leaves lots of opportunity for other specialists to do their work and among them they get the whole job done. (See Chapter 3 for another discussion of niche.) Other hypotheses have proposed that more species originate in and adapt to warm climates than cool climates or that species go extinct more rapidly in cold environments than warm ones. A recent genetic study of birds and mammals in the Americas suggests that both evolution and extinction are faster in the high latitudes than in the tropics. Arctic birds and mammals, though fewer in number, are younger than those in the species-rich rainforests of Central and South America, perhaps a product of the rapid climate changes of the Pleistocene Epoch. The key questions remain: Why are there so many species anywhere? And where did they all originate?

Evolutionary Patterns

Evolution by natural selection is the underpinning of modern biology and biogeography, including the concept of a biome. Biomes are composed of species possessing adaptations to a particular set of environmental conditions. An adaptation is a genetically controlled trait that is the product of evolution, enhances the survival and reproductive success of individuals, and has become a characteristic of the species as a whole. Adaptations may be clearly visible aspects of plants and animals such as size, shape, and color; or they may involve metabolic processes related to biochemistry, timing of metabolic activity, and other aspects of physiology. In animals, genetically controlled (instinctive) behaviors are adaptations to either the physical or biotic environment.

Plants

The most obvious adaptations in plants are elements of morphology or physiognomy. These include the size and shape of the entire plant; the size and shape of its leaves; characteristics of stems and branches including woody versus nonwoody, smooth barked or rough barked, thorny, swollen, and so forth; and the habit or pattern of leaf replacement, whether foliage remains all year (evergreen habit) or all leaves are shed for the nongrowing season (deciduous habit). Two of the most commonly used systems for categorizing plants according to their physiognomy are those devised by Christen Raunkiaer (1860–1938), a plant geographer from Denmark, and Pierre Dansereau (1911–), a Canadian ecologist and biogeographer.

....

Evolution by Natural Selection

Charles Darwin (1809–1882) and Alfred Russel Wallace were co-founders of the theory of evolution by natural selection. Natural selection is the means by which *species* adapt to their environment and is simply a matter of differential reproductive success. Those individuals whose traits best let them withstand the adverse or restrictive aspects of their environment—be they living or nonliving factors—and best exploit its resources will usually survive more often and breed more successfully than will individuals less well equipped or less "fit." The more fit individuals will leave more offspring bearing their traits in the next generation, which thus will contain a larger proportion of individuals with the better-suited traits. The trend will continue from generation to generation until there is a significant change in the environment. In this manner, nature selects the members of the species that contribute the most genetic information to future generations and the species becomes better adapted to the conditions in which it lives.

....

Raunkiaer identified what he called lifeforms on the basis of the location of perennating or renewal bud, the tissues from which new growth develops each year (see Table 2.2 and Figure 2.3). Some of his terminology—geophyte, hydrophyte, and epiphyte—remains in common usage as descriptors of plant types.

Dansereau based his categories on the overall physiognomy of plants and called each type a different growthform (see Figure 2.4). He first distinguished among trees, shrubs, herbs, lianas (woody vines), epiphytes, and bryophytes—the nonvascular mosses, liverworts, and lichens. The woody plants, the trees and shrubs, are further differentiated according to leaf shape, size, texture, and habit. Herbs are split between forbs—broadleaved, nonwoody plants, and graminoids—plants with grass-like blades, including the grasses themselves.

Leaves may be thin and flat and broad (broadleaves), or they may be narrow and needle-like (needleleaves). The term needleleaf is usually restricted to describing the foliage of conifers (gymnosperms). Leaves of all angiosperms are considered broadleaves, even though they may be reduced to small size and narrow shapes or thickened and leathery, or succulent. Each variation is viewed as an adaptation to a particular set of environmental conditions.

Both needleleaf and broadleaf trees and shrubs may be either evergreen or deciduous. Most needleleaf trees are evergreen, but deciduous types such as the larches and the bald cypress do exist. Broadleaf trees and shrubs are usually evergreen in warm humid climate regions where precipitation is abundant all year, as in the Tropical Rainforest Biome. They are usually deciduous in regions with strong seasonality of either temperature patterns or precipitation patterns, as in the Tropical Seasonal Forest Biome and the Temperate Broadleaf Deciduous Forest Biome. Desert shrubs are also frequently deciduous.

Forbs are those plants we often call wildflowers or weeds. They may be annuals, completing their life cycles in a single year; they may be biennials, requiring two years to germinate, grow, flower, and set seed before dying; or they may be perennials regenerating their aboveground parts year after year from the same subsurface bulb, corm, or rhizome—the geophyte lifeform of Raunkiaer.

Graminoids include sedges and rushes as well as grasses. These represent three different families of flowering plants: Cypercaeae, Juncaceae, and Graminae, respectively.

Table 2.2 Raunkiaer's Lifeforms

Phanerophytes	Tall woody plants (trees, large shrubs, and vines): renewal buds form more than 10 in (25 cm) above the ground.
Chamaephytes	Small woody plants or herbaceous shrubs with stems that persist from year to year: renewal buds form less than 10 in (25 cm) above the ground.
Hemicryptophytes	Small perennial plants with stems that die back at the end of the growing season: renewal buds form at ground level or just below the surface.
Cryptophytes	Small perennial plants with bulbs, corms, or rhizomes: renewal organs form well below the surface.
Geophytes	Terrestial cryptophytes.
Helophytes	Aquatic plants such as emergents in lakes: renewal buds are on rhizomes buried in the waterlogged lake bottom.
Hydrophytes	Aquatic plants with submerged or floating leaves: renewal buds held in the water column.
Therophytes	Annual plants: renewed from seeds.
Epiphytes	Plants that grow perched on the branches and stems of larger plants and have their roots dangling in the air.

Danscreau's scheme is easier to apply, especially in the field, and easier to visualize than Raunkiaer's. It has become the more common way to describe the types of plants that characterize the different biomes, but often some of Raunkiaer's terms are applied and the two systems blend together.

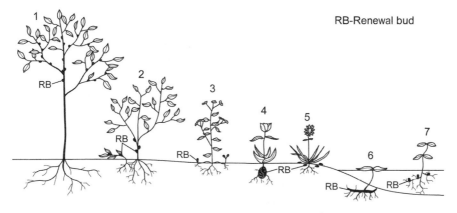

Figure 2.3 Raunkiaer's Lifeforms. This system of categorizing plants according to form rather than species is based on the position of the renewal or perennating bud. 1 = phanerophyte; 2 = chamaephyte; 3 = hemicryptophyte; 4 = geophyte; 5 = therophyte (only reproduces from seed); 6 = helophyte; 7 = hydrophyte. *(Illustration by Jeff Dixon.)*

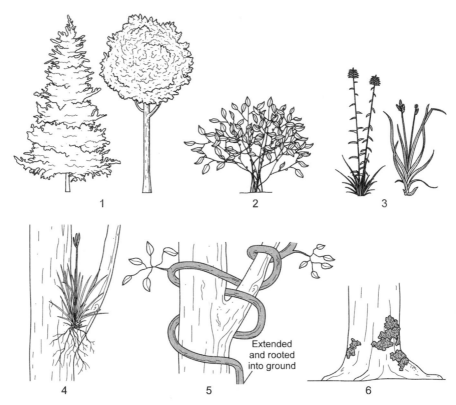

Figure 2.4 Dansereau's growthforms are based on the structure or appearance of plants. 1 = tree (left = needleleaf, right = broadleaf); 2 = shrub; 3 = herbs (left = forb, right = graminoid); 4 = epiphyte; 5 = liana; 6 = bryophytes. *(Illustration by Jeff Dixon.)*

Vegetation

The most prevalent growthforms found at a particular area and how they are positioned on the ground surface determine the structure of the vegetation, the arrangement of photosynthetic surfaces vertically from ground to canopy and horizontally according to the space between plants. Structure relates to layering or stratification of foliage because of the co-occurrence of plants of different heights. Cover measures the percentage of the ground shielded from direct sunlight by the each layer of foliage.

Six structural vegetation types are usually recognized: forest, woodland, savanna, grassland, scrub or shrubland, and desert or desertscrub. Forests have trees as the dominant growthform and a closed canopy layer. A closed canopy is a tree layer in which the crowns of neighboring trees overlap and prevent most sunlight from reaching the forest floor, either all year in the case of evergreen trees or during the growing season in the case of deciduous trees. Forests may have a complex vertical structure since short and tall trees, shrubs, forbs, bryophytes, epiphytes, and lianas may all grow beneath or in the canopy.

Woodlands are also composed of trees, but the canopy is more open than in forests and sufficient light reaches the ground to let forbs, graminoids, and bryophytes grow as a dense, though sometimes patchy, surface layer. The vertical structure usually is simpler, with only two or three layers.

Savannas are a structural vegetation type characterized by a continuous groundcover of grasses with an open canopy of trees or shrubs above it. The woody plants may be widely scattered or clumped together in patches. The resulting parklike landscape is most closely associated with the Tropical Savanna Biome, but oak savannas characterize some mediterranean areas.

Grasslands lack woody plants, except perhaps along streambanks, where they occur as linear gallery forests. The dominant growthforms are graminoids and forbs. In the Temperate Grassland Biomes the white, yellow, pink, and blue flowers of perennial forbs may be visual dominants in spring and summer.

Shrublands are common elements today in regions with a mediterranean climate. Rainfall occurs during the cool winter months and the summer growing season is typically dry. This precipitation pattern imposes water stresses on the plants. Shrubs are better able to tolerate severe drought than grasses and trees. Nonetheless, considerable debate exists about whether the shrublands of mediterranean areas result from climatic controls or are the product of long human use and abuse of the land. It may be that woodlands or savannas would cover large parts of these regions were it not for clearing for agriculture, prolonged browsing by goats and other livestock, or repeated burning.

Deserts are dominated by the shrub growthform. The vegetation type may be referred to as desertscrub, and the term desert used to describe the arid climate. Deserts are not wastelands, as they are too frequently depicted. Shrubs are often widely spaced, but hidden in the soil waiting for rain are geophytes or perennial forbs and the seeds of annual forbs. In hot deserts yet another growthform, the succulent, may be conspicuous. Succulents may store water in leaves, stems, or roots. Several plant families, including aloe, cactus, euphorb, lily, mesembrantheum, and milkweed, have succulent members thriving in the world's hot deserts.

Animals

Animals adapt to climatic conditions, to the plant members of the biome, and to other animals to fulfill their requirements for food, shelter, and mates. Latitudinal gradients in bird and mammal morphology were recognized in the nineteenth century and formulated into ecogeographic rules (see Figure 2.5). Bergman's Rule stated that the colder the climate—that is, the higher the latitude, the larger the body mass of species compared with close relatives living under warmer conditions and at lower latitudes. Christian Bergmann (1814–1865), for whom the rule is named, explained the phenomena by noting that large bodies (volumes) have relatively less surface area per unit of weight than small bodies. Since body heat is radiated to the environment through the skin, the lowered ratio helps large animals maintain body heat in cold regions. Conversely, mammals and birds in warm

Figure 2.5. These three foxes compared with one another display elements of all three ecogeographic rules: (left) The arctic fox (10–20 lbs/4.5–9 kg) has the largest body mass of the three and lives at the highest latitudes, reflecting Bergmann's Rule. Its ears are small and snout relatively short, as Allen's Rule predicts. *(Photo by A. DeGange, USFWS.)* (center) The red fox is intermediate in size (4–15 lbs/1.8–6.75 kg) and latitudinal distribution. An animal of humid climates, it is much darker than the desert-dwelling kit fox, confirmation of Gloger's Rule. *(Photo by R. Laubenstein, USFWS.)* (right) The San Joaquin kit fox is a warm desert dweller. Its large ears follow Allen's Rule, while its small size (3–6 lbs/1.5–2.7 kg) is anticipated by Bergmann's Rule. Its relatively pale color is expected according to Gloger's Rule. *(Photo by B. Peterson, USFWS.)*

climates regions are faced with the problem of getting rid of excess heat. A large ratio of surface area to volume facilitates cooling and helps maintain core body temperatures.

Joel Asaph Allen (1838–1921) added to this concept by recognizing that mammals and birds in cold regions had proportionately shorter limbs, and in the case of mammals shorter ears and snouts, than relatives in warmer latitudes. This is known as Allen's Rule. These adaptations further reduce surface area and heat loss in polar regions. Longer limbs increase surface area and promote heat loss in tropical and desert regions.

A third ecogeographic rule is Gloger's Rule, named after the Polish zoologist Constantin Gloger (1803–1859), who noted geographic variations in body coloration (see Figure 2.5). Gloger's Rule recognizes that warm-blooded animals in more humid environments tend to have darker pigments in skin, hair, or feathers than animals in drier climate regions. This may be related to the need to absorb more ultraviolet wavelengths to produce vitamin D in the tropical forests. Dark brown and black pigments may increase resistance to bacteria that infect skin and feathers. The converse, of course, is that the drier the climate the lighter is the color of the plumage or pellage. Light colors reflect more sunlight than dark pigments and may help in the regulation of body temperatures in hot deserts. They also may camouflage individuals against the tans and grays of the desert backdrop.

In addition to the traits mentioned in the ecogeographic rules, other morphological adaptations confer fitness in particular habitats. For a life in the trees, an arboreal life, prehensile tails are common among the mammals of the Neotropical rainforests and act as a fifth limb with which to grasp the branches. This form is exemplified in the New World monkeys. Handy as this is, no African or Asian monkey possesses a grasping tail. Asian apes such as gibbons and orangutans travel through the rainforest using their long arms to swing from branch to branch. Other Old World forms run and jump from branch to branch like squirrels. There is no one way to do it. Success can come from a variety of adaptations.

Among river-dwelling animals that must struggle to maintain their position in fast-moving water, stream-lined bodies or small flat bodies are common successful adaptations. Long legs ending in hooves are advantageous to large mammals in grasslands that must run away from predators or undertake long migrations. The bodies of invertebrate predators and vertebrates have been shaped in innumerable ways to make them efficient ambushers or chasers of prey. Similar adaptations are often found among animals of different taxonomic relationships that are part of the same biome, products of what is known as convergent evolution.

Physiological adaptations in animals fine-tune them to their environments. These involve metabolic mechanisms that allow animals to conserve body moisture in desert environments, or undergo periods of dormancy (hibernation) in cold climate regions, or schedule their reproductive cycles to coincide birthing with periods of maximum food availability.

Behavioral adaptations are common among animals and vary from biome to biome as one would expect. They may represent the fastest way species can evolve and adjust to changes in environmental conditions. Some behavior is learned, but most is instinctive; that is, it is under genetic control and the product of natural selection. One example of a behavioral adaptation is the activity period. Some species are nocturnal, active only at night. (Morphological adaptations such as large eyes may assist their nighttime searches for food.) Others are diurnal, active by day and safely hidden away at night. Some desert animals find they are most successful if they are active at dusk and at dawn, when temperatures are lowest and the humidity relatively high. These crepuscular animals seek shelter both at night and during the daytime.

In open vegetation, such as that of savannas and grasslands, large mammals often gather together in herds. Many ears and eyes are then alert to threats posed by large predators. These might seem advantageous to desert animals also, but sparse vegetation and limited water render large groups a nonviable option. No one adaptation, be it morphological, physiological, or behavioral, works alone. All traits evolve within a matrix of competing needs and the best solutions are usually compromises.

Further Readings

Internet Sources
The following Web sites provide access to biographic information on many of the pioneers of biogeography and ecology, as well as to abstracts and some full articles of classic works.

Smith, Charles H. 2002–2008. "Early Classics in Biogeography, Distribution, and Diversity Studies: To 1950." Western Kentucky University. http://www.wku.edu/%7Esmithch/biogeog.

Smith, Charles H. 2003–2008. "Early Classics in Biogeography, Distribution, and Diversity Studies: 1951–1975." Western Kentucky University. http://www.wku.edu/%7Esmithch/biogeog/index2.htm.

3

Ecological Concepts Important to the Study of Biomes

The idea of a biome traces back to naturalists such as Buffon, but the term itself and many refinements of the concept are the work of ecologists. Ecology is the science that studies the interrelationships of species and their environment in its totality, including both nonliving and living elements. It is a young science; and as it developed in the late nineteenth and first half of the twentieth century, it contributed new understandings of how nature worked that were then incorporated into the study and mapping of biomes. This chapter introduces some of the major concepts and terminology from ecology that are key to the modern view of biomes.

Ecosystems

The basic unit of study in ecology is the ecosystem. This is defined as a community of species and the physical environment with which it interacts to maintain a flow of energy and a cycling of nutrients. An ecosystem is tied to a particular area and so can be mapped. Yet the boundaries are usually vague and the size or scale varies according to the interests and needs of the researcher. An ecosystem can be as small as a 4 in (10 cm) sealed glass orb, an "EcoSphere," in which microorganisms—tiny shrimp and algae—form a self-sustaining saltwater system that usually lasts from six months to two years. Or the Earth itself (called Gaia in this conceptualization) can be considered an ecosystem, one where life has been sustained for 2.5 billion years, albeit in vastly different communities at different times in the geologic past. The largest regional or continental-scale ecosystem, recognizable by the similar vegetation and climate throughout, is a biome.

Ecosystems perform functions and have structure or organization. These are really two sides of the same coin, inseparable since structure reflects function. All ecosystems receive inputs of energy, usually but not always sunlight, and then transfer it through the community, using and degrading some at each step along the way. And all ecosystems transfer matter (nutrients, but also toxins) from species to species and from the living to the nonliving and back again. The two major functions of any ecosystem, therefore, are to maintain a flow of energy and to recycle nutrients. A number of other vital functions can be categorized as services. Services include pollination of plants, dispersal of seeds, regulation of population sizes, and the promotion of disturbances that lead to renewal of the community.

Energy Flow

Energy flow in most ecosystems is initiated with the absorption of solar radiation by photosynthetic cyanobacteria, algae, bryophytes, or higher plants. Photosynthesis transforms the light energy to chemical energy that becomes stored in the bonds that hold organic molecules together. The chemical energy can either be used directly by the photosynthesizing organisms themselves for their own metabolic processes or passed on to other organisms. Since photosynthetic cells and organisms prepare usable forms of energy for the entire community, they are called the primary producers or simply the producers in the ecosystem. (In a few extreme environments such as at deep sea hydrothermal vents and cold seeps, no photosynthesis occurs. The producers are *chemosynthetic* bacteria that transform chemical energy from sulfur compounds or methane into those kinds of chemical energy that can be used by other forms of life.)

The energy not used by photosynthetic organisms is transferred to others when they are consumed or when they die and decay. Energy follows one of two pathways through the ecosystem, although in truth the two are intertwined (see Figure 3.1). A grazing pathway passes the energy from plants to plant-eating animals, the herbivores or primary or first-level consumers. Carnivores or animal-eating animals receive what energy has not been used by herbivores when they kill and eat a grazing animal. Carnivores are sometimes called secondary or second-level consumers. Animals that consume carnivores are called top carnivores or third-level consumers. It is rare in a land-based ecosystem for enough energy to remain and be stored in top carnivores for a population of animals feeding nearly exclusively on top carnivores to be sustained. However, in aquatic ecosystems in which energy flows start with cyanobacteria and algal cells and in which first-level consumers are tiny zooplankters, there may be a few more levels of consumers.

The grazing pathway for energy flow creates a grazing food chain, an organization or structure based on what eats what. A food chain is the simplest representation of this pathway. In all but the simplest of ecosystems, a food web in which different species consume the same food is closer to reality. Each species has

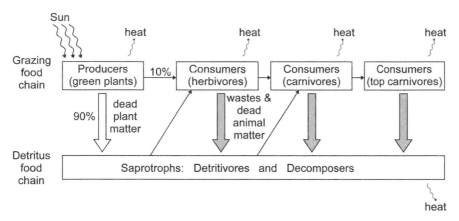

Figure 3.1 Energy flows through a terrestrial ecosystem by first being fixed in a usable, chemical form by the producers (green plants), which use some for their own metabolic processes and store the remainder as new plant tissue (growth and reproduction). That which is used is converted to heat energy and released to the environment. The stored energy is available to consumers via the grazing food chain. If not consumed, it passes to the detritus food chain. About 10 percent of the energy fixed by plants (primary production) flows through the grazing food chain and 90 percent through the detritus food chain. At each step energy is used and converted to heat energy, which exits the system. Energy from the sun must be constantly resupplied since energy cannot be recycled. *(Illustration by Jeff Dixon.)*

evolved to take its energy from a particular position along the flow or path where it can most efficiently capture it. This position is the species' niche. Niche is an abstract concept, but can be thought of as the way a species obtains its food or how it exploits its habitat. Habitat is the physical place an organism inhabits and often is described either by site conditions or by vegetation type.

Not all energy entering the grazing food chain stays in it. That which is used by living organisms is transformed into heat energy and dissipates into the environment. Not all plants or all parts of plants are consumed by herbivores. Uneaten plant tissues eventually die, but energy remains locked in the bonds of organic molecules. Herbivores ingest plant material that they cannot digest and regurgitate or defecate it as waste products with the bonds still intact. Metabolic processes generate other wastes. The same story goes for carnivores and top carnivores. Not all herbivores are caught and eaten, and even among those that are caught not all parts are consumed and not all parts that are ingested are digested. Carnivores are a bit more efficient at extracting energy than are herbivores, but at best only 10 percent of the energy they consume is available to be passed on to the next level of consumers. The rest of the energy is either metabolized or eliminated as feces. Carcasses of carnivores and top carnivores represent energy not going into the grazing food chain. Along with the dead plant matter, dead herbivores, and all the animal wastes, this energy goes into a detritrus food chain (see Figure 3.1).

The breakdown of detritus and release of its bound energy and transfer to living organisms begin with saprotrophs: decomposers and scavengers. Saprotrophs include bacteria, fungi (especially molds), and some protists that can extract energy from dead and decaying plant and animal matter. Decomposers are those saprotrophs that in the process of extracting energy from organic matter reduce it to inorganic molecules, releasing carbon dioxide to the atmosphere and preparing nutrients for reuse by the plant community. The main decomposers are fungi. Scavengers are organisms that eat carrion, animals already dead when first discovered. The boundary between hunter and scavenger, however, can be blurred at times and some carnivores both hunt and scavenge. Scavengers help prepare dead organic materials for decomposition, but do not themselves decompose it. They are best considered detritivores, members of the group of organisms that feed on detritus. The indigestible materials they leave behind—lignin, bones, hair, and the like—will be processed by the true decomposers.

Grazing and detritus food chains or webs are one way to represent the structure of an ecosystem. Another common depiction is based on the energy flow through the grazing food chain but represents it as a pyramid of trophic (feeding) levels (see Figure 3.2). The graphic display of the concept gives substance to the abstract notion of levels. The primary producers form the base of the pyramid and make up the largest level whether measured in biomass or numbers of organisms. Each higher level gets smaller as the energy available from the level below diminishes. While detritivores and omnivores, which feed at more than one level, are neglected, this standard model is a simple way to demonstrate the interplay between energy, ecosystem structure, biomass, and total numbers of organisms.

Aquatic ecosystems follow the same general patterns as terrestrial ones with the addition of one food chain and one or more levels on the trophic pyramid. The additional food chain is the microbial loop, in which organic matter and the energy in it move from the phytoplankters in the form of dissolved organic matter (DOM) leaking from their cells to bacteria, which are then fed upon by protists and other tiny zooplankters (see Figure 3.3). The bacteria decompose the DOM and reduce it to inorganic nutrients that can be taken up by the phytoplankters. The rapid cell

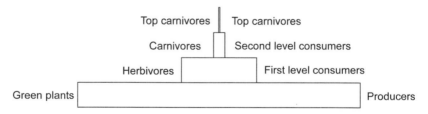

Figure 3.2 The trophic pyramid is one representation of the structure of the ecosystem. The length of each bar may represent the biomass, the number of individuals, or the amount of energy in that level. The producers are always the largest level in terrestrial ecosystems. *(Illustration by Jeff Dixon.)*

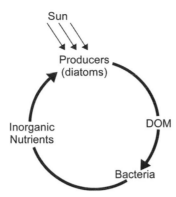

Figure 3.3 The microbial loop is important in the flow of energy and the cycling of nutrients in aquatic biomes. Phytoplankters (algae such as diatoms) are the chief primary producers. Their cells leak dissolved organic matter (DOM), which feeds bacteria that break down the organic compounds into their inorganic components. The inorganic compounds serve as nutrients for algae, completing the cycle. *(Illustration by Jeff Dixon.)*

division of phytoplankters supports the zooplankton, which is fed upon by larger invertebrates, which are fed upon by small fishes, which in turn are consumed by larger fishes, which may be caught by fish eagles or seabirds. Together these account for at least six trophic levels in contrast to the four typical in terrestrial ecosystems.

Nutrient Cycles

Because energy is carried through the ecosystem in organic matter, it is bound to nutrients and follows the nutrient cycles. The difference is that energy is degraded and lost to the system as it is used; matter can be reused and recycled. The basic nutrient cycle traces the change from inorganic compounds absorbed by plants to organic compounds produced during photosynthesis and metabolic processes back to inorganic compounds by processes of organic decomposition. The carbon cycle (see Figure 3.4) best illustrates this since carbon is the main building block of living (or organic) material. Complex compounds of carbon are manufactured by living organisms, becoming what are known as organic compounds. Carbon is held in inorganic form as carbon dioxide in the atmosphere. It is taken up by green plants and converted to complex sugars during photosynthesis. Chemical energy holds the molecule together. When the plant needs to use energy to live, it releases the energy in the chemical bonds through the process of respiration. The organic molecules combine with oxygen, which breaks apart the sugar and frees the energy and also releases carbon in the form of carbon dioxide. The cycle is completed.

Other nutrients similarly change back and forth from inorganic to organic forms. Several different nutrients are necessary for life processes to occur. Perhaps

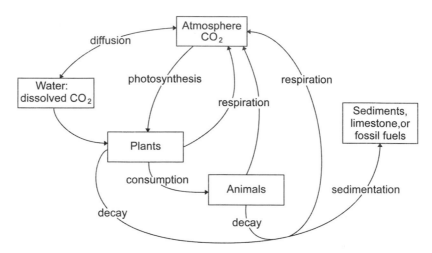

Figure 3.4 The carbon cycle demonstrates the circular pathway that elements take between inorganic forms abundant in the atmosphere or bedrock and the organic compounds formed by living organisms. Respiration and decay break organic compounds back into their inorganic forms. Sedimentation or incomplete decay leading to fossil fuels can trap elements in long-term storage. *(Illustration by Jeff Dixon.)*

most important are the plant nutrients nitrogen and phosphorus, although calcium, sodium, magnesium, and others are also vital. Nitrogen has as its main reservoir the main component of the Earth's atmosphere, nitrogen gas. However, the triple bonds of nitrogen gas (N_2) cannot be broken by most organisms. It must first be altered or "fixed" into a usable form that dissolves in water. Specialized nitrogen-fixing bacteria and cyanobacteria accomplish this by forming ammonia (NH_3), nitrite (NO_2^{-1}), and—most important—nitrate (NO_3^{-1}). Dissolved nitrates are absorbed through the roots of plants or are absorbed through the cell walls of algae and converted to such essential organic compounds as enzymes, proteins, and deoxyribonucleic acid (DNA). These compounds will be reprocessed into animal enzymes, proteins, and DNA as they are transferred through the food chains. Finally, a different group of bacteria will reduce these large molecules back to inorganic compounds when they decompose dead organic material. Soluble nitrates are readily washed out of the soil, so nitrogen is frequently a limiting factor in plant growth.

   Phosphorus, calcium, and other nutrients mentioned above differ from nitrogen and carbon in that they derive from bedrock and not from atmospheric gases. Phosphorus occurs in nature as phosphates (PO_4^{-3}). An essential nutrient, phosphorus is used in the transfer of energy through plant and animal cells and organisms, which is accomplished in the high-energy bonds of adenosine triphosphate (ATP). Phosphates are frequently applied as crop fertilizers, and runoff from agricultural fields is a major source of nutrient pollution in lakes and estuaries. Calcium also comes from bedrock, usually in the form of calcium carbonate, which in water disassociates into the positive calcium ion (Ca^{+2}) and negative bicarbonate (HCO_3^{-1}) or carbonate (CO_3^{-2}) ions. Calcium and other positively charged ions or bases, such as sodium (Na^{+1}) and magnesium (Mg^{+1}), play significant roles in internal water regulation and in nerve impulse generation. A common measure of soil fertility is the base ion exchange rate, and part of the importance of humus in soils is its ability to trap these bases and prevent their being leached to depths beyond the reach of roots.

Ecosystem Services

All the functions grouped together as services promote the long-term survival of the system and its component species. Populations must reproduce. Species must be able to disperse to new sites or recolonize old sites after communities are damaged or destroyed in order to survive. One species must be held in check so as not to out-compete others for resources or wipe out prey species. Aging systems need to be revitalized with new inputs of nutrients or opened canopies to let in more sunlight. Microclimates and microhabitats must be conserved; the water table must be kept from dropping too low; pollution of groundwater or shallow littoral waters must be prevented. Ecosystem services are recognized as services to human communities as well as. Crops need to be pollinated, floods and storm-wave damage minimized, erosion prevented, and potable water supplies maintained.

The significance of ecosystem services is evident in the number of adaptations related to them. Among the more obvious is the coevolution of some plants and their pollinators. Sometimes only one insect has the correct size and shape of proboscis to reach nectar hidden deep in the flowers of a particular plant species. The plant may open its blooms only when that insect is active and may have special colors and scents that attract its sole pollinator. Dispersal of seeds is essential for the survival of most plants, and many species have evolved ways to attract animals to come and carry their fruits or seeds away. Barbs help some seeds stick like Velcro to passing animals. Other plants produce sweet succulent fruits so that they will be eaten and the seeds will be passed safely through the digestive tracts of birds or bats and other mammals to be deposited in nutrient-rich feces at some distance from the parent plant. Fire is necessary for some plants if their seeds are to germinate. Shaded under the canopy of dense shrubs or trees or stifled by dense grass cover, the seedlings of sun-loving plants may be doomed. But if a fire removes the canopy, sun streams in and they may grow quickly to reclaim their species' hold on the territory. Some shrubs in California's expression of the Mediterranean Biome seem to promote repeated burning by containing flammable oils. Browsing and grazing may accomplish the same end as fire in some tropical savannas and temperate grasslands. Studies reveal that grazed prairies maintain more plant species than ungrazed ones. Destruction of trees by elephants in African woodlands and seasonal forests opens the canopy and lets grasses grow, inviting a large number of ungulates to coexist in a mosaic of habitats at various stages of damage and recovery.

Populations of one species may also create habitat for others. This may be most clearly seen in marine environments, where the hard shells of mussels, for example, provide attachment sites for other sessile organisms such as barnacles and hydroids. Between the shells, fine particles become trapped and provide habitat for burrowing animals such as polychaete worms. Without the mussels, the ecosystem would have many fewer species circulating nutrients and building intricate food webs to promote the flow of energy. In forests, cavity-nesting birds would be absent if it weren't for the trees. In rainforests, epiphytes use tree branches and lianas for support. Juvenile fish and molting crabs need seagrass meadows to hide in. These are just a few examples of one species providing living quarters for others.

Ecosystem services are important to the study of biomes because they help determine the distribution, composition, vulnerability, and longevity of plant and animal communities.

Ecosystem Development: Ecological Succession

The complex interactions among the physical environment and the living community of microbes, plants, and animals that are the hallmark of an ecosystem develop over time. As new species come to occupy a site, they alter the physical and biological conditions. Nutrient cycles change, food chains change, and vegetation or habitat structure changes. The new set of attributes may favor colonization by other species and become disadvantageous to the current resident species of the area, which then disappear. This was the original explanation for changes observed in the species composition of a particular patch of habitat over time and the replacement of one community by another. The process of change was called ecological succession or simply succession. Today, ecosystem change is not viewed in so straightforward a manner, but the descriptive value and vocabulary of ecological succession remain useful.

History of the Concept

Although certainly not the first to remark on vegetation change, University of Chicago plant ecologist Henry Chandler Cowles (1869–1939) was the first to formalize the concept of plant succession. Trained in both botany and geology, Cowles studied the vegetation on the Indiana Dunes on the shores of Lake Michigan. He noted a clear spatial pattern in different types of plant life occupying the ever older and more stabilized sand dunes that occurred at progressively greater distances from the lake's modern beaches. Close to the lake shifting dunes supported wiry beach grasses. Inland on ancient shorelines the dunes were stable. Stable dunes closest to the lake were covered in shrubs; at intermediate distances cottonwoods and pines grew. Then came a ring of dunes covered with oaks and hickories. And finally the oldest dunes were masked by the regional temperate broadleaf deciduous forest dominated by beeches and maples. Stressing the connections between vegetation and geology, Cowles proposed that vegetation on sand dunes of different ages represented different stages in the general trend of vegetation development over time on dunes. Each dune and its plant cover represented one step in the sequence or sere of plant communities that would occupy the area, beginning with dunes devoid of vegetation and ending with them part of the broad regional vegetation. Cowles' doctoral dissertation, "The Ecological Relations of the Vegetation of the Sand Dunes of Lake Michigan," was published in 1899 and became a classic in the literature of ecology.

Cowles thought that successive plant communities tended toward a stable equilibrium, but that a final unchanging assemblage of species was never actually attained. Species came and went, even though they were linked together in symbiotic types of relationships. Succession did not proceed in a straight line. There were setbacks, cycles, and side trails along the way.

Frederic Clements, the same botanist who coined the term biome, had a different view of succession, and his view came to dominate ecology for many years. He studied the grasslands and coniferous forests of the northern prairies and saw these vegetation types as stable and concrete expressions of the regional climate. For him, succession was directional (see Figure 3.5). It proceeded with the orderly occupation of a barren site by pioneering plants, followed by a predictable series of different plant communities, and ended with a final permanent or climax community determined by the climate. The climax was discernible by the growthform of the dominant plants. Since all stages leading to the development of the climax would eventually give way to climax vegetation, he identified the climax as the potential vegetation of the region and equated it with the idea of a biome.

Clements thought of plant communities as large organisms. All parts were interconnected and the survival of one member depended on the survival of all members. Communities had life histories through which they grew, matured, and died. And they had evolutionary histories in which they adapted, accepted, or rejected innovation and became better tuned to the climatic conditions they faced. Clements devised a whole taxonomy of plant communities to identify various stages of development; most of these terms are no longer used. His notion of a stable climax in balance with the climate also has been largely abandoned. Yet the

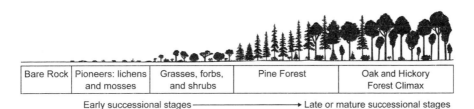

| Bare Rock | Pioneers: lichens and mosses | Grasses, forbs, and shrubs | Pine Forest | Oak and Hickory Forest Climax |

Early successional stages ⟶ Late or mature successional stages

Figure 3.5 The classic model of ecological succession. *(Illustration by Jeff Dixon. Adapted from Odum 1959.)*

general concept of ecological succession proceeding in identifiable stages and the terms that imply a maturing ecosystem are still used (see below).

One of the most influential challenges to Clement's way of thinking came from Henry A. Gleason (1882–1975). Gleason championed an "individualistic" concept as opposed to the Clementsian organismic view. His observations led him to propose that species operated independently of each other. Each had its own evolutionary and dispersal history, its own tolerances and requirements, and only happenstance and similar adaptations caused a number of species to have overlapping distributions and occupy the same place at the same time. Each assemblage of plants was the unique result of individual species' responses to fluctuating physical conditions, even those occurring on short-term time scales. Survival of one species did not depend on the presence of all others. Some could be removed and the community or ecosystem would survive.

Gleason's view gained widespread acceptance as the better way to explain how nature works. Ecology now focuses on change as a universal characteristic of ecosystems. Nonequilibrium communities are the norm. Nevertheless, ecologists still use Clementsian terms and speak of early or mature successional communities and even climax (and subclimax) vegetation to describe the mosaic of plant associations that make up a given landscape or biome. The change is not believed to be predetermined, but the various stages he identified are evident, nonetheless. The model of ecological succession remains a useful introduction to ecosystem change.

The Classic Model of Primary Ecological Succession

The basic model of primary ecological succession considers only plant life and starts with a site with no life. This could be a new glacial lake, a lava flow, a sand dune, or the ground exposed as a glacier retreats. Bare ground or a lifeless body of water will soon be colonized by a variety of plants that begin to establish nutrient cycles, shade the ground, contribute to the weathering of rock, or otherwise alter the environment. As the environment changes, new species colonize and displace some of the pioneers that are not as well adapted to the new conditions. Over time, a predictable series of plant communities will replace one another on the site until a final community—the climax community—in balance with the regional climate takes over and persists indefinitely. A key element of succession is that the series of communities develop in orderly fashion. Different groups of plants form the early stages according to the original substrate (water, unconsolidated, sands, or solid rock), but with time they all proceed to the same climax in a given climate region.

On solid bedrock, for example, the bare surface receives intense sunlight and temperature fluctuations can be extreme. Rainwater runs off immediately, so as plant habitat the rock is dry. No soil exists in which flowering plants can become rooted; and without plants and the humus derived from them, nutrients released from the bedrock are rapidly washed away. This is a harsh environment regardless of climate and only a few organisms are able to colonize it. The first to come, the pioneers, are often crustose lichens.

Lichens immediately alter the physical environment. They shade patches of rock and moderate surface temperatures somewhat. They release acids that chemically fragment bits of solid rock to smaller particles. They hold moisture on the surface and they concentrate nutrients. Other forms of lichens—foliose, fruticose, and squamulose—gain a foothold and continue to create a habitat increasingly favorable for moisture-loving bryophytes such as mosses. Mosses and dead plant matter provide a rooting medium for higher plants. Grasses and forbs appear, especially those species harboring symbiotic rhizobial bacteria that can fix nitrogen. The rock continues to weather, plants contribute humus, decomposers recycle nutrients, and soil formation begins. Climate permitting, grasses and forbs may be replaced by low shrubs. Members of the heath family seem particularly well suited to the still-harsh conditions and shallow soils of this yet early successional stage. They may shade out the grasses and forbs. With time the soil deepens, more humus forms and retains nutrients, and larger plants can invade. In early maturity a tree community of short-lived nitrogen-fixing birches or aspens may dominate, but with continued development of an ever-richer soil and with more moderate temperatures and increased humidity caused by the greater shading of the ground, these small trees will be replaced by other, longer-lived species, such as the oaks and maples that make up the canopies of mature temperate deciduous forests. Under a mild continental or a humid subtropical climate, this forest will endure for millennia—until a major change in climate occurs. It is the climax community.

The series of communities developed in predictable sequence on a dry site is called a xerosere. Hydroseres are the series of communities that begin with open water and a lake bottom devoid of life. Phytoplankters are the first to colonize and begin fixing energy and cycling nutrients. Simultaneously, sediments are being washed in from the land. This material not only will be rich in nutrients, but also will provide a substrate in which aquatic plants such as pondweeds can become rooted. Roots bind the loose particles together and dead plants add organic matter to a thickening sediment layer. As the water thereby becomes shallower, emergents such as cattails and sedges become established. Enriched water stimulates phytoplankton production and the growth of rooted plants, which trap more sediments and build up the lake bed around its margins. The area of deep open water diminishes and the pond eventually becomes a marsh. As the soil builds up above the water table, drainage improves and the substrate dries out. The marsh may give way to a meadow. The meadow will be invaded by shrubs and small trees, and later the trees of the climax community.

At any given time, different parts of a region host communities in different stages of development or succession. Thus, a patchwork of vegetation actually covers large stretches of territory in the same climate region. In some cases, successional communities (or seral stages) that are not climax persist for long periods of time, prevented from continuing on to the climax stage by repeated disturbance or edaphic conditions. Frequent fire, for example, is considered by some to prevent tropical savannas from turning into climate-determined tropical dry forests.

Similarly, mediterranean scrub vegetation might proceed into an elfin forest or a savanna with oak trees were it not for fires occurring every 10–35 years. In the boreal forest, patches of muskeg are a characteristic part of the vegetation mosaic. These bogs are self-perpetuating features of the landscape, maintained long term as waterlogged habitats by the capacity of sphagnum moss to hold water and by the acidic humus of sphagnum mosses that keeps nutrients levels too low for most other kinds of plants to become established. Such long-lived plant communities not "in balance" with the climate are known as subclimaxes.

Secondary Succession

The classic or Clementsian view of succession allowed for the readily observable change in vegetation that happens after the abandonment of agricultural land or after a forest fire. Under these circumstances and others in which plants once occupied a site but disturbance of one kind or another greatly damaged or destroyed it, a soil is already in place. The starting point for succession is thus different than on a site that had never hosted life, and the pioneering species are not the same. Succession occurring after disturbance is called secondary succession, and typically it proceeds much more rapidly than primary succession. Abandoned cropland in the humid subtropical climate used above for examples of primary succession would be invaded first by weeds. The term weed refers to any species adapted to rapid dispersal and fast colonization of open spaces. Seeds are often small and produced in great numbers. The wind blows them helter-skelter so some are bound to land on open ground. They are sun-loving plants and germinate and grow quickly in unshaded locations. Most are annuals or biennials, completing their life cycles in a brief period, ready to move on when larger plants invade the site and cast their shadows. Many are forbs, but some trees have "weedy" or opportunistic tendencies. After a few years of dominance by crabgrasses, asters, ragweeds, and broomsedge, pines invade. In a few more years, the pines grow tall enough to shade out the grasses and forbs, but they too are sun-loving plants and do not create conditions favorable for their own regeneration. Instead of pines perpetuating themselves on the site, they allow shade-tolerant broadleaf trees to come in. As these oaks and maples and beeches grow large, an understory of other shade-tolerant plants develops and species-rich herb and ground layers thrive. The climax community of the regional forest is once again in place.

The Contemporary View of Succession

Students of vegetation change no longer accept a predictable and long-lasting outcome to succession. The idea of a permanently unchanging climax community belongs to another era. Much more credence is given to chance and timing and to the separate responses of each species to environmental change than was allowed for in the classic model. Site conditions and the interactions of species already occupying the site are important, but so are the dispersal capabilities of individual species. The first to arrive may simply be those most easily dispersed or those with nearby populations (sources of colonists) intact. Which way the wind is blowing may determine which seeds are blown onto the site and which are not. Late arrivals may or may not be influenced by site conditions. They may simply be slow dispersers or come from distant locations.

That plant species migrate individually rather than as a whole community has been revealed in studies of pollen dating to the late Pleistocene (see Chapter 4). Scientists have been able to trace the movement of broadleaf trees out of refugia in southern Europe and the southern United States at the end of the last Ice Age. As climate warmed, one by one, species advanced northward and formed what we think of today as the vegetation of Temperate Broadleaf Deciduous Forest Biome. Some species (of animals, as well as plants) are still on the move. In addition to these natural migrations, other species have been introduced to continents far from their place of origin by humans. These exotic species also reshape ecosystems, making what exists today different from anything that existed in the past. Such additions (and subtractions, because extinctions also occurred) raise the question, what is pristine? Or for that matter, what is natural?

Disturbance is a regular feature of ecosystems, and therefore of biomes. Some are resilient and bounce back to more or less original states. It is most improbable that exactly the same species will be recovered, however, since exactly the same set of interacting living and nonliving conditions will no longer occur. Sometimes disturbance will so greatly alter the community that an entirely new ecosystem will become established.

Disturbance usually occurs too frequently for a true climax community to be a possibility. Cycles of vegetation destruction and recovery or replacement may be a more appropriate model of the real world. Cycles can be maintained by nutrient depletion of soils (as seems to happen in some old-age coniferous forests), by regular droughts, periodic flooding, hurricanes and ice storms, lightning-set fires, overgrazing, or any number of other naturally recurring events.

Thinking in terms of cycles rather than a straight-line succession of communities with an endpoint changes the way one thinks about nature. Forest fires usually do not "destroy" acres and acres of land, as frequently reported in the media. They reset or renew the cycle of vegetation change. A sudden input of nutrients released from the vegetation when it burns stimulates the growth of forbs and other sun-loving plants. They contribute humus to the soil and prepare the way for reinvasion by trees. Life goes on. Constant change also means that the old idea of a "balance of nature" is just that, an old idea. It suggests a static situation that would fall apart if a single element were removed. The new dynamic view of nature sees interchangeable parts and built-in redundancy, so if one element is lost another is available to take on its role in the ecosystem. Resiliency rather than stability is the sign of a healthy ecosystem. And most ecosystems are resilient, at least up to a point. If too much change happens too rapidly, they can collapse. What will replace them is a matter of what the site conditions are like, which species are available to colonize, and what dispersal routes connect source populations to the newly vacant site.

Soil

The development of a soil has been linked to ecological succession and comes about from the interplay of climate, vegetation, and bedrock. Soil is a complex mixture of varying proportions of mineral particles, decaying organic matter, microorganisms, and living and dead plants and animals, water solutions, and gases. Good definitions elude even soil scientists who may be quick to say it is not the same thing as dirt, but struggle to say exactly what it is beyond a thin surface layer capable of supplying and storing plant nutrients. Zonal soils—those that are associated with specific climates and vegetation types—show distinct layers or horizons that develop through one of three major soil-forming processes. Most biomes have a characteristic zonal soil, so the following discussion emphasizes the major zonal soils and ways they are believed to form.

A soil begins to develop when bedrock exposed to the atmosphere begins to disintegrate or weather into finer and finer particles. This action contributes the mineral or inorganic fraction of the soil, which constitutes the bulk of a mature soil. Loose particles of weathered bedrock do not make a soil but rather a material called regolith. Other deposits of mineral particles may also be the starting point of soil formation. Sand, alluvium, loess, and glacial deposits may become parent materials for soils. The major mineral components of soils are compounds of silica, iron, and aluminum.

Temperature and moisture determine the rates at which chemical and biological processes work to alter the parent material, increasing them in warm, humid climates and slowing them in cold or dry climates. Water percolating through the regolith dissolves some minerals and removes ions and fine particles from the upper parts of the soil column and washes them into lower sections. This leaching of soluble materials alters the chemical composition of the regolith and produces distinct layers. In the wet tropics, chemical weathering occurs rapidly and leaching removes soluble plant nutrients. Since these regions were not bulldozed bare by continental ice sheets during the Pleistocene Epoch, many tropical surfaces have been exposed to rain and high temperatures for long periods of time measured in millions of years. Deep, nutrient-poor soils are the result. In the high latitudes of the Northern Hemisphere, old soils and regoliths were scraped away by the ice and only a few thousand years have been available for new soils to form. Cold winters further limit the time during which most weathering processes can operate, so soils tend to be thin and rocky, though many young soils are rich in nutrients.

Plants have many essential roles in the development of zonal soils. The roots offer paths for water to drain through the ground and help atmospheric gases vital to life penetrate to depth. The main functions, however, are in concentrating nutrients and maintaining nutrient cycles. The root hairs of plants, often with the help of mycorrhizal fungi, draw in dissolved nutrients and prevent their being leached out of the soil column. The nutrients are then assimilated into the tissues of the plant, where they are stored until leaves are shed or the plant dies. Dead leaves, twigs, and other plant matter accumulate on the ground as litter. As the litter decays, its organic compounds are converted back to inorganic mineral nutrients. Rainwater and snowmelt infiltrate and transport the minerals down into the soil, where they can be trapped once again and recycled by new root systems.

Bacteria and fungi are biological agents of decay that process dead plant and animal matter. As such, they are of vital importance in nutrient cycles and the perpetuation of life on the Earth, even though they are often overlooked. Other microorganisms in the soil have equally important functions in maintaining nutrient cycles. Free-living cyanobacteria photosynthesize and add organic materials. Nitrogen-fixing bacteria render nitrogen gas from the atmosphere into the soluble nitrates that plants can use. Other bacteria break down nitrogenous organic wastes into ammonia and nitrites and then into nitrates at the other end of the nitrogen cycle.

Those bacteria and fungi that process dead matter produce a material that is an intermediate stage between recognizable plant tissue and inorganic mineral, namely, humus. This slick black substance is composed of colloidal-size organic particles. Together with similarly sized clay particles produced by the weathering of the parent material, humus attracts and holds ions that otherwise might be leached down beyond the reach of root hairs (see Figure 3.6). Clay and humus are negatively charged flat platelets. Many important plant nutrients, such as calcium, magnesium, and potassium, are positively charged ions or cations and so are drawn to the humus-clay particles. The humus and clay bind the cations strongly enough to prevent their being washed away but loosely enough that they can be given over to plants' roots. The scientific measure of soil fertility determines the cation-exchange capacity of the soil. An easier way to estimate fertility is to examine the color of the soil. If it is dark brown or black, it has a lot of humus with many cations likely being conserved for plant use. Humus also helps hold water in the soil.

A key plant nutrient, nitrate (NO_3^{-1}), is a negative ion and not attracted to or conserved by humus and clay particles. Nitrates are easily leached from soils, and for this reason, nitrates are often a limiting factor in plant growth. Nitrates and ammonia, another source of nitrogen, are common fertilizers applied by farmers to

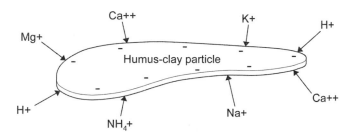

Figure 3.6 A humus-clay particle. The negative charges on these colloid-size particles attract positively charged ions and help hold them in the A horizon of soils. Many of these ions are important plant nutrients. Under acidic conditions, however, the more reactive H+ ions can swamp the humus, causing soils to have low natural fertility. *(Illustration by Jeff Dixon.)*

their crops. Today these fertilizers are mostly artificially produced and are major components of nutrient pollution in lakes, rivers, estuaries, and shallow bays.

Soil water is important in dissolving nutrients and preparing them for uptake by plants. It also can leach the ions to lower depths if plants do not immediately use them or if there is insufficient humus to hold the ions temporarily in the upper levels. Plant roots also need oxygen, so some air must be available in the spaces between soil particles. Waterlogged soils are detrimental to the growth of many plants unless they have evolved special structures to serve as breathing tubes. Too little water is also a problem causing plants to wilt and depriving them of essential dissolved nutrients. Yet evaporation of soil moisture can be beneficial as it draws soil moisture and its dissolved substances back up to the root level. If the upper parts of the soil contain little water, some soil moisture can rise by capillary action through the air-filled pores. Upward movement of soil moisture is important in the development of dryland soils.

Soil Fauna

Both invertebrates and vertebrates are important elements in soil development, and many spend all or parts of their lives below ground. Some contribute to the detritus food chains and act as decomposers. Others, both those living below the surface and those living on the surface, contribute nutrients via their waste products and dead bodies. Small animals that burrow and tunnel in the soil mix particles and aerate the soil. They carry some materials down and move some upward. Their holes and passages are routes by which water and air can penetrate to depth.

Earthworms are one example of soil animals that enhance soil fertility by returning nutrients to the root zone as they churn through the soil. In the tropics, termites and ants provide this ecosystem service. In the temperate grasslands, burrowing rodents such as prairie dogs (North American prairies) and mole rats (Eurasian steppes) are natural plows that turn over the soil.

Soil Profiles

As soils develop over time, they deepen and the materials in them become altered and rearranged. Organic matter is added; weathering and decay convert inorganic and organic matter, respectively; dissolved materials and colloids are leached out or at least carried to lower levels. The visible result in a mature soil is a distinct set of vertical layers or horizons (see Plate XV). A slice through the ground that reveals the layers in the soil column is called a soil profile (see Figure 3.7). In its simplest form, a soil profile consists of three horizons—A, B, and C—although as many as six horizons are recognized by some authors.

On the surface, particularly in forests, is an organic horizon (sometimes referred to as the O horizon) consisting of litter and humus. In other biomes, the surface horizon is the A horizon or topsoil, usually rich in organic matter such as dead roots and humus. The topsoil is where most seeds germinate. The lower part of the A horizon is a zone from which colloids and soluble materials have been

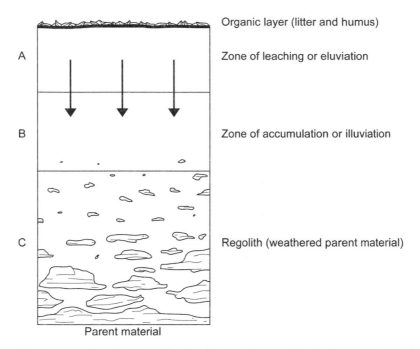

A — Organic layer (litter and humus)

Zone of leaching or eluviation

B — Zone of accumulation or illuviation

C — Regolith (weathered parent material)

Parent material

Figure 3.7 A generalized soil profile showing the three major horizons that develop in mature soils from interactions among climate, vegetation, and the parent material. *(Illustration by Jeff Dixon.)*

removed. The process by which percolating water carries fine particles downward through the soil column is called eluviation, so this zone of removal is frequently identified as a separate E horizon.

The B horizon is a zone of accumulation or illuviation, where clays, iron oxides, and aluminum oxides removed from the A horizon are deposited. This is the subsoil and is often heavier and more clayey than the A horizon. Under some climate regimes calcium carbonates precipitate out in this horizon. Since clays may trap nutrient cations, plows are designed to reach this level and turn it so as to bring it and its nutrients to the surface as a way of enriching the root zone.

The C horizon consists of regolith, the unconsolidated parent material of soils. Weathering occurs in this horizon and continues to supply the soil with fresh inorganic material. Below the regolith is the bedrock or other unweathered parent material.

Although the development of a clear profile is considered a sign of a true, mature soil, this usually happens only in well-drained, undisturbed sites. Many soils are too young or too old and leached to exhibit distinct horizons. So-called young or immature soils lack a B horizon. Additional or repeated horizons may occur where soils have developed under one or more past climatic regimes and fossil horizons remain.

Soil-Forming Processes

Climate and vegetation ultimately overcome the influences of bedrock and determine the nature of a soil in a given region. Although the soil-forming processes and end products vary gradually across geographic space—as does climate—three basic soil-forming processes are recognized: laterization, podzolization, and calcification (see Figure 3.8). Each results in the development of horizons with distinct properties and becomes dominant in different biomes.

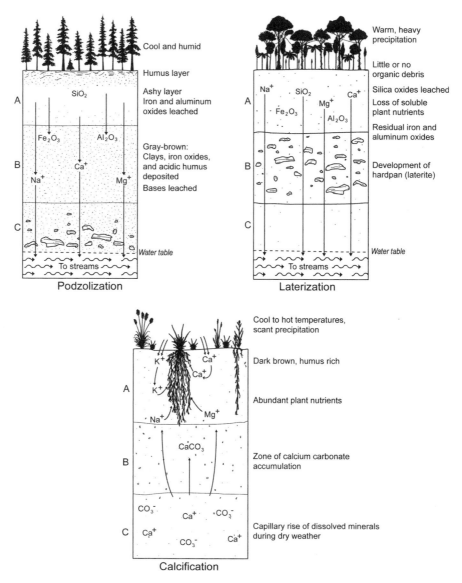

Figure 3.8 The three major soil-forming processes. Each is associated with a particular vegetation type and climate regime. *(Illustration by Jeff Dixon.)*

Laterization. In warm humid areas, the most common soil-forming process is laterization. It is associated with ancient surfaces in the tropics and produces the typical low-nutrient soils underlying many tropical rainforests and tropical savannas. In laterization, weathering of the bedrock is extreme as is the leaching of soluble materials. Under the year-round warm temperatures of the tropics and high amounts of rainfall each year, the silica compounds in the regolith and soil are "mobilized," that is they become soluble and are leached from the soil column along with the nutrients. What is left in the soil are compounds of iron and aluminum. Iron oxides (ferric oxide, Fe_2O_3) impart a deep red or rust color to most well-drained tropical soils. Poorly drained soils are yellow from a different form of iron oxide (ferrous oxide, FeO), which forms in anaerobic environments. In some areas, aluminum oxide concentrations are so great that the soil is mined for bauxite, the ore of aluminum metal.

Humus is lacking in the soil even beneath lush rainforests. Two factors combine to make this so. First, the forests are evergreen. Although the leaves do not live forever, dead leaves drop one by one all through the year. At any given time, the layer of litter on the ground is sparse. The second factor relates to the decomposition of the litter into humus. With high temperatures and moist conditions year-round, bacteria and fungi are active all of the time. The rate of decay is so rapid that litter passes quickly through the humus stage to separate into its inorganic components. Rainforest trees typically have shallow root systems to trap as many of these inorganic ions as possible before they are leached away.

Podzolization. In some ways podzolization can be thought of as the opposite of laterization. The process takes place under needleleaf forests and is therefore most closely associated with the Boreal Forest Biome, although podzolization can occur even in the tropics. Water passing through the litter of needleleaf plants produces an acidic soil moisture. Chemically an acid is a substance that disassociates into a hydrogen ion (H^+) and some negative ion. Hydrogen ions are highly reactive and attracted to humus and clay particles, on which they displace nutrient cations such as calcium. The released nutrients become vulnerable to leaching and are removed from the upper levels of the soil. Under acidic conditions, aluminum and iron oxides are mobilized and also leached from the A horizon. Left behind is an obvious layer of gray powdery silica oxide. The process derives its name from this ash-like material, podzol—the Russian word for ashy. (Russian scientists pioneered the study of soils and their name for this widespread process has been accepted around the world.)

The substances washed from the A horizon are redeposited in the B horizon, which acquires a reddish brown color from humus and iron compounds. The resulting soils are of limited fertility, but the main reason they are not widely utilized for agriculture relates to the short growing season at high latitudes and not to soil quality.

Calcification. Under semiarid and arid climates, excessive leaching is not a consideration and nutrients tend to concentrate in the soils because of the contrasting movements of soil water via percolation and capillary action. When water percolates through the A horizon, calcium carbonate dissolves. In the B horizon, as the water evaporates, the ions bind together again and calcium carbonate precipitates out. Sometimes the presence of calcium carbonate can be determined only by squirting acid onto an exposed part of the horizon and observing that it fizzes as acid and base react. The drier the climate (that is, the more potential evapotranspiration exceeds precipitation) the more visible the calcium carbonate becomes either as white nodules or as layers of light gray hardpan known as caliche.

The calcification process is characteristic in the Temperate Grassland Biome, where it produces some of the world's most fertile soils. The perennial grasses play a key role as their fine deep roots let scarce water penetrate to depth and then rise back through the B horizon. The seasonal die-down of grasses and the die-off of many roots each year adds abundant humus to the soil. Grassland soils are usually black or dark brown and rich in plant nutrients.

At the other extreme in the arid lands of the world, in deserts receiving less than 10 in (250 mm) of precipitation a year, little humus forms because the plant cover is sparse and dryness inhibits decay processes. Calcium carbonate is concentrated in the soil close to the surface and the entire soil column is whitish or gray in color. These soils, when brought under irrigation, can be quite fertile since they contain many undissolved nutrients. The threat of salinization of irrigated land is great, however; high evaporation rates can produce a glistening crust of salts on the surface if water is not properly managed.

Soil Types: U.S. Soil Taxonomy
American soil scientists classify soils in a hierarchy of categories resembling the Linnaean taxonomic hierarchy used for living organisms. The soil "taxon" that most directly correlates with biomes is the soil order. Each order ends in the suffix "-sol" (for soil) and bears a descriptive root term. Orders contain soils with similar characteristics of horizon development and degree of weathering. Each order contains numerous local soil species that are not relevant to a global discussion of biomes.

Entisols. Soils in this order are usually composed of young parent materials only recently exposed to weathering ("Ent-" comes from recent). Entisols thus have poorly developed profiles. Prevalent in dry areas where soil-forming processes are slow, entisols are not restricted to any particular biome. Many have formed on sands.

Vertisols. These soils have a large clay component that causes them to swell and shrink as seasons alternate between wet and dry. The back and forth of expansion and shrinking churns the upper levels and inhibits profile development. ("Vert-" comes from the Latin *verto*, meaning to turn.) The black cracking clays under some African and Australian tropical savannas are examples.

Andisols. Andisols derive from recent volcanic deposits. Because they are young, the soil profile is poorly developed. However, because they are of volcanic origin, they are high in plant nutrients. The name comes from the Andes Mountains, where volcanic eruptions continue to spread volcanic ash. Andisols are found in other volcanic areas as well, including Japan, Indonesia, and the Palouse regions of Washington, Oregon, and Idaho.

Inceptisols. Another group of immature soils with weakly developed profiles are the inceptisols. ("Incept-" comes from the Latin *incipere*, to begin.) What characteristics do faintly distinguish the horizons are mostly the result of the removal of the finest particles from the A horizon. They are primarily associated with the Tundra Biome and with mountains.

Aridosols. Desert soils, aridisols develop where water is insufficient to leach soluble minerals. ("Arid-" means dry.) They lack organic content since little vegetation is available to produce litter and humus. Hence most are sandy, alkaline, and light colored. Aridosols are widespread in the Desert Biome and estimated to cover 20 percent of Earth's surface.

Mollisols. The deep black and brown soils characteristic of the Temperate Grassland Biome are mollisols. ("Moll-" comes from the Latin *mollis*, meaning soft.) The texture and color of good potting soil, the deep A and B horizons have an abundance of clay-humus particles and retain plant nutrients. The richest of these soils developed with calcium-rich loess as the parent material and still bear the Russian name for black soils, chernozem. These soils probably have the highest natural fertility. Their use in agriculture was limited before the nineteenth century because no plows had yet been invented that could cut through the thick sod of the world's steppes, prairies, and pampas. John Deere's self-scouring steel plow, first manufactured in 1838, began the conversion of grasslands to wheatfields and cornfields.

Spodosols. Spodosols are the product of extreme podzolization ("Spodo-" derives from the word *podzol*). These soils are associated with the Boreal Forest Biome and subarctic climate, but develop under pine forests or other acidic situations in any climate, even in the tropics. The whitish A horizon is diagnostic. It is leached of iron and aluminum compounds as well as humus and nutrient cations. The horizons are all acidic, so fertility is relatively low.

Alfisols. Once known as gray-brown forest soils, alfisols are produced by soil-forming processes intermediate between true podzolization and true laterization. ("Alfi-" pertains to the high proportion of aluminum, Al, and iron, Fe, left in the A horizon.) These soils are associated with cooler parts of the Temperate Broadleaf Deciduous Forest Biome, but they are not restricted to these regions. Developing under broadleaf deciduous forests, they receive a seasonal pulse of litter that is much less acidic than that which forms under coniferous forests. With longer, warmer summers than experienced in the boreal forest, decay of the litter is more rapid and humus that is able to capture and hold nutrient cations is abundant in both A and B horizons. Until mollisols could be cultivated, the alfisols were the preferred agricultural soils in Europe and North America, where the forest was relatively easily cleared, even with primitive tools.

Ultisols. Warmer regions of the Temperate Broadleaf Deciduous Biome, especially in the southeastern United States and China, where greater weathering and more complete leaching of nutrients from the soil have occurred than farther north, ultisols develop a reddish color caused by the concentration of iron and aluminum oxides in the A horizon. Humus content is low as a result of faster decay rates brought on by the longer periods of warmer temperatures and the higher amounts of rainfall in humid subtropical climates compared with the warm summer continental climates. In the north-south continuum from podzolization to laterization in humid climates, these soils lie closer to laterization than do alfisols. Although fertility is lower than that of alfisols, these soils were used to grow tobacco and cotton in the Old South. Through poor agricultural practices, much of the topsoil was lost, and today it is mostly the deep red clay subsoil that remains.

Oxisols. The exposed continental shields of the humid tropics host the most weathered and leached soils of all, the oxisols. The product of extreme laterization, they were formerly called latosols. Deep red or yellow horizons composed primarily of iron and aluminum oxides are typical ("Oxi-" refers to the dominance of oxides), but the separation between the A and B horizons is barely perceptible. They lack humus and plant nutrients. Despite the low natural fertility of these soils, lush tropical rainforests grow on them because the plants are able to take up cations as soon as they are released by decomposition processes or have developed other strategies to obtain nutrients from stem flow or out of the air. The vegetation, not the soil, holds the nutrients. When the forest is cleared, the ecosystem's nutrients are quickly depleted. Oxisols are associated with the Tropical Rainforest Biome, the Tropical Seasonal Forest Biome, and the Tropical Savanna Biome.

Histosols. A group of soils that does not really fit the criteria for a true soil, histosols are purely organic in composition. (The prefix "histo-" refers to body tissue, hence organic material.) Typically they are saturated all or most of the time. As waterlogged wetland soils, they lack oxygen and therefore aerobic bacteria. Decay proceeds slowly, so much dead plant material accumulates as peat or muck. Most histosols are black and acidic. They are largely found in glaciated areas of the middle and high latitudes, but can be found in swamps well to the south of the limits of Pleistocene glaciation.

Further Readings

Books

See any ecology textbook for other descriptions and explanations of processes mentioned in this chapter. Consult any physical geography textbook for information on soils and soil formation.

Real, Leslie A., and James H. Brown. 1991. *Foundations of Ecology. Classic Papers with Commentaries.* Chicago: University of Chicago Press. A selection of key articles or excerpts from the major works of early ecologists, including Cowles, Clements, and Gleason. Covers a period from the late-nineteenth century to 1970.

Internet Sources

McDaniel, Paul. 2007. "The 12 Soil Orders, Soil Taxonomy." University of Idaho, College of Agricultural and Life Sciences, Soil and Land Resources Division. http://soils.ag .uidaho.edu/soilorders/index.htm.

U.S. Department of Agriculture, Natural Resources and Conservation Service. n.d. "Distribution Maps of Dominant Soil Orders." http://soils.usda.gov/technical/classification/ orders. Photos, brief but technical descriptions, and U.S. soil maps can be accessed from this site.

4

Major Environmental Factors in Terrestrial Biomes

The distribution of terrestrial biomes is largely controlled by atmospheric processes that influence the distribution of heat and moisture and geological processes that shape and reshape the Earth. The atmosphere and oceans are linked in a system of heat transfer and moisture cycling that varies in detail with latitude and determines climate. The size and geographic position of the continents, the occurrence of major mountain ranges, and the location of cold and warm boundary currents in the seas all modify the basic pattern. Climate in turn affects the rates at which bedrock wears down and soils form. Interactions among air, sea, land, and living organisms result in the development of a distinct vegetation type characteristic of each climate region, the essence of a biome. The patterns we see today represent a freeze-action snapshot of ongoing processes on a dynamic planet. The atmosphere is changing; ocean floors are spreading; continents are moving; some mountains are wearing down and others are still building up. Knowledge of aspects of the geologic past help one understand the present and predict the future. This chapter examines the ever-changing factors of the physical environment that have major influences on the distributional pattern of the world's land-based biomes and the organisms that inhabit them.

Latitude

Latitude measures distance north or south of the Equator and is the key factor in the climatic differences found around the world. Latitude is expressed in degrees, minutes, and seconds because this distance is measured as an angle with its origin

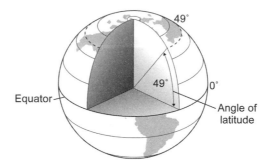

Figure 4.1 Latitude is an angular measure of distance from the Equator. *(Illustration by Jeff Dixon.)*

at the center of the planet. One side stretches to the Equator, the other to the location in question (see Figure 4.1).

Latitude matters in climate because the Earth is essentially a sphere and therefore the intensity of sunlight received at any point on its surface depends on where that point is (and the time of year) (see Figure 4.2). Since all of the energy involved

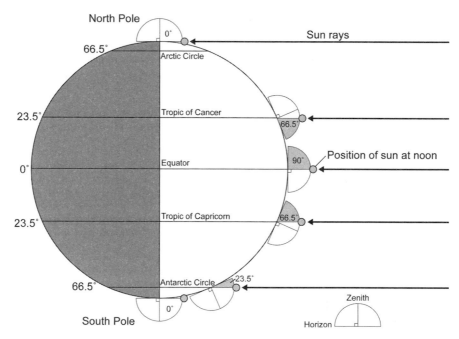

Figure 4.2 The intensity of solar radiation received at any place on the Earth is a function of the angle at which the sun's rays strike the surface. The curved surface of the Earth means that vertical or direct rays are received at only one latitude on a given day. Everywhere else the sun's rays come in at an angle less than 90 degrees. The tilt of the axis determines exactly which latitude receives direct rays at which time of year. *(Illustration by Jeff Dixon.)*

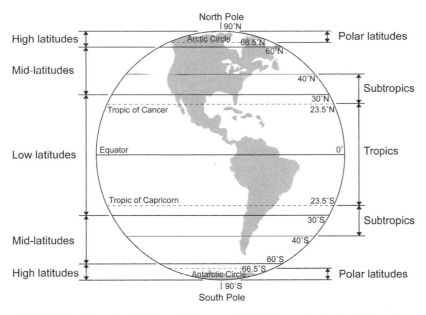

Figure 4.3 Major latitudinal belts and key parallels. *(Illustration by Jeff Dixon.)*

in atmospheric processes (and life processes) ultimately derives from solar radiation, the amount of energy received is important. The differing amounts of energy received at different latitudes cause differences in surface heating that set the atmosphere into motion and result in a global system of pressure and winds. The moving atmosphere carries heat poleward from the Equator, drives the ocean currents, and transports moisture from sea to land.

When considering biomes at a global scale, exact latitudes are rarely needed. Instead, it is useful to think in terms of broad latitudinal zones or belts (see Figure 4.3). Nonetheless, a few latitudes and their parallels are important to know because they indicate extremes of solar radiation, either seasonally or annually. The Equator at 0° latitude is the parallel that receives the direct or vertical rays of the sun twice a year, at each equinox. The Tropic of Cancer (23° 30′ N) and Tropic of Capricorn (23° 30′ S) mark the farthest distances from the Equator ever reached by the vertical rays and define the tropics. The Arctic Circle (66° 30′ N) and Antarctic Circle (66° 30′ S) are the most equatorward parallels that experience 24 hours of daylight or 24 hours of darkness (depending on the season) at the solstices.

Seasons

Seasonal variation in atmospheric conditions occurs because the Earth revolves around the sun on a tilted axis. The axis is always inclined in the same direction, and the North Pole points to the North Star or Polaris (although this has not

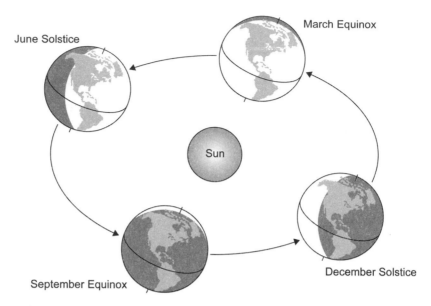

Figure 4.4 Earth-sun relationships change as the Earth revolves around the sun, because the axis always points in the same direction—to Polaris, the North or Pole Star. The result is differing daylengths throughout the year and changes in the angle of incoming solar rays. Four specific positions in the orbit were identified even in ancient times and became the basis of calendars: the two equinoxes and the two solstices. *(Illustration by Jeff Dixon.)*

always been the case), so that in different points along its orbit, Earth's poles change orientation relative to the sun (see Figure 4.4). Therefore, the point at which the vertical rays of the noon sun strike the surface—the subsolar point—changes its latitude from day to day. People have long recognized four points along the orbit— the equinoxes and solstices—and have based their annual calendars upon them. At an equinox, the vertical rays fall on the Equator. By the June solstice, they have shifted north 23.5° to fall on the Tropic of Cancer. The tilt of the axis (23.5°) determines that this position is as far north as the vertical rays ever reach. After the solstice, the position at which the vertical rays strikes migrates back to and across the Equator, arriving at the Tropic of Capricorn on the December solstice.

Coinciding with the position of the vertical rays, though with a lag time, is the time of greatest heating of the atmosphere. When the sun is in the Northern Hemisphere, as it is often expressed, it is summer in the Northern Hemisphere. In the middle latitudes, long daylengths and warm to hot temperatures are the hallmarks of the season. Polar regions are bathed in low-angled sunlight all day long. In the tropics, the rain-bringing equatorial low and Intertropical Convergence Zone (ITCZ—see below) have moved across the area, and it is the wet season.

The opposite is happening in the Southern Hemisphere. At the June solstice, the subsolar point is as far from the Southern Hemisphere as it will get. It is winter

in the Southern Hemisphere, where in the mid-latitude days are short and temperatures are cool. Because continents do not extend to high latitudes as they do north
of the Equator and because much less land overall occurs south of the Equator than
to the north, winter temperatures are not as extreme as in, say, Canada or Russia.
The Antarctic region is plunged into 24 hours of darkness. The tropics beyond the
equatorial latitudes (that is, from 10° to 23.5° S) experience their dry season.

Temperature Controls

Temperature is primarily a function of latitude and the basic Earth-sun relationships that determine how much and when insolation is received during the year.
These two factors produce the main spatial variations in surface heating as well as
seasonal temperature differences. Temperature is a measure of heat energy and
actually indicates how fast atoms are vibrating in a substance. The obvious geographic trend is from warmer temperatures at the Equator to colder temperatures
at the poles. Various factors modify this simple pattern somewhat.

Elevation is a second important control of temperature. Everyone is familiar
with the fact that mountaintops are cooler than surrounding lowlands. This in part
is because of the lower density of the atmosphere at high altitudes. Fewer gas molecules and more widely separated molecules result in fewer collisions to stimulate
vibration. Temperature decreases with increasing elevation at variable rates, but
the average is 3.6° F per 1,000 ft (6.5° C per 1,000 m), a vertical temperature
change known as the normal lapse rate.

Land and water respond differently to incoming solar radiation and have different heating and cooling rates. Land is opaque and heat concentrates at the surface.
Temperature changes between day and night and between summer and winter are
rapid and can be extreme. This is a phenomenon known as continentality. Oceans
on the other hand respond slowly. Water is transparent and sunlight can penetrate
to some depth and become diffused through the medium. In addition, the peculiar
chemistry of water requires a lot of energy to be absorbed and held as latent energy
before the polar bonds that link water molecules can be weakened enough to allow
increased vibration (see also Chapter 5). Water stores heat energy in the latent form
and releases it only slowly as the sensible heat that thermometers measure, that we
feel, and by which the atmosphere above is warmed. Large bodies of water warm
slowly and cool slowly and prevent extreme temperatures on nearby land areas, an
effect known as maritime influence.

Ice and snow keep temperatures low because they reflect so much of the sun's
light energy. Melting ice caps are evidence of warming trends, but their shrinkage
also contributes to global warming as more and more solar energy is absorbed by
bare ground and converted to heat energy.

Water vapor in the air is a greenhouse gas that is unevenly distributed geographically and temporally. Its effects are most noticeable on humid and cloudy

nights, when it traps heat radiating from the ground and prevents significant cooling of the surface. Cloudiness depresses daytime temperatures, because sunlight is reflected back to space at the top of the clouds and is scattered as it passes through the cloud layer. Since the equatorial regions are usually cloud covered, their temperatures do not get as high as those of subtropical deserts, where little impedes the incoming rays.

Ocean currents also affect temperatures of coastal areas. Warm currents transport heat from the tropics toward the poles and raise temperatures above what might be expected for the latitude. Cold currents offshore depress temperatures along tropical coasts.

Global Patterns of Pressure and Winds

The air above the Earth circulates between the surface and the lower atmosphere (see Figure 4.5), limited by gravity from escaping to outer space. The year-long nearly direct rays of sunlight received on the ground near the Equator result in

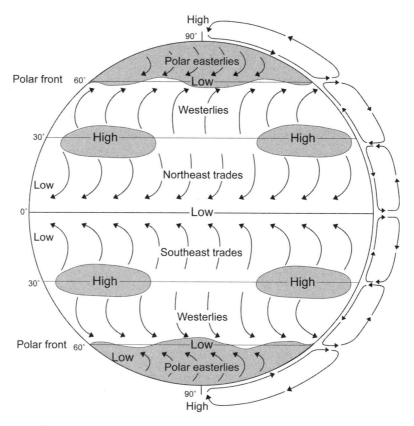

Figure 4.5 Global model of semipermanent atmospheric pressure systems and prevailing winds. The latitudinal positions of both pressure belts and winds shift with the seasons. *(Illustration by Jeff Dixon.)*

intense heating of the air above, which causes it to expand, become less dense than surrounding air, and rise.

The relatively low density of the atmosphere in equatorial latitudes results in a global belt of low atmospheric pressure that shifts during the year between roughly 10° N and 10° S. The rising air cools and as a consequence becomes less and less able to hold water vapor. The excess water vapor condenses and clouds form. With continued rising and cooling, precipitation occurs (see also the section Precipitation Controls). Very warm equatorial airmasses are initially able to hold large amounts of water vapor. As this moisture is expelled from rising, cooling air, rainfall amounts greater than 100 in (2,500 mm) a year become common.

The rotation of the Earth forces the air that rises in equatorial latitudes back down to the surface around 30° of latitude in both the Northern and Southern hemispheres. As the air descends, it becomes dense and it warms. As a result, essentially permanent zones of high atmospheric pressure form in the subtropics. The continents break up the belts so that concentrated *cells* of high pressure over the oceans dominate instead of a single world-girdling zone. Since the airmass is warmer when it reaches the ground than it was high in the atmosphere, it can now hold more moisture than it contains. Water will be drawn from the oceans through the process of evaporation until the airmass becomes saturated. However, lifting of the air to cool it becomes unlikely in a zone in which strong downdrafts occur, so arid climates and deserts are common in subtropical latitudes.

The air that comes down to the surface in the subtropics flows out from the centers of high pressure. Some moves equatorward, drawn into the equatorial low. There the air from both hemispheres meets to rise again at what is known as the ITCZ. The remainder flows poleward toward low-pressure zones near 60° latitude in both hemispheres. As this air moves across the surface of a rotating planet, it is deflected by the Coriolis Force and the air spirals out of the highs like giant pinwheels. The Coriolis Force causes air (and water) to be deflected to the right of its intended path in the Northern Hemisphere and to left in the Southern Hemisphere. This pattern becomes especially significant in determining the yearly or seasonal aridity of subtropical west coasts and the development of cold ocean currents offshore.

Other regions dominated by high pressure occur at the poles, where very cold, very dense air descends from above. These polar highs are, like the subtropical ones, source areas for air (winds) flowing across the surface. They are also dry regions, since little opportunity exists for the uplift needed to release moisture, and the very cold air cannot hold much water vapor.

Cold air moving equatorward from polar highs meets warmer airmasses moving poleward out of the subtropical highs at about 60° N and S. The different temperatures mean different airmass densities, and at the contact zone, the warmer, less dense air rises over the colder, denser mass. This confrontation between airmasses of different temperatures and different densities was once viewed as a battleground between advancing "armies" and termed a front. Specifically, the zone of rising air at 60° is known as the Polar Front. As a place where air is rising, the Polar Front is a zone of low pressure and is associated with storminess and heavy amounts of rainfall, though not as much precipitation falls as at the equatorial low because the air is cooler and holds less water vapor overall. Weather at latitudes influenced by the Polar Front is changeable from day to day. When polar airmasses are temporarily winning the day at a particular place, the weather is sunny and humidity low. When the advancing subtropical air "army" wins the battle, the day is cloudy and drizzly.

Air flowing across Earth's surface is wind. Winds blow from high-pressure areas to low-pressure areas. Thus, the global system of pressure dictates a global system of winds modified by the Coriolis Force. Winds are named according to the direction from which they come. Polar Easterlies dominate between 60° and 90° latitude; Prevailing Westerlies blow between 60° and 30° latitude; and the steady easterly Trade Winds characterize those parts of Earth between the Equator and 30° latitude (see Figure 4.5).

Mountain ranges can divert winds or force the air to rise over them. When air rises over mountains the so-called orographic effect results. Rising air cools and releases its moisture on the windward side of the range, creating higher-than-normal amounts of precipitation. As the air flows down the opposite or leeward side, it warms. Precipitation stops and drier-than-normal conditions (for the latitude) develop what is known as the rainshadow effect (see Figure 4.6). Many mid-latitude deserts lie in the rainshadow of a high mountain chain. The temperate rainforests of the Americas are products of orographically generated rainfall.

Heating and cooling of the mid-latitude regions of the continents can also alter wind directions. The most important product of such seasonal changes in temperature is the Asian monsoon. A monsoon is a wind that reverses its direction with the seasons. Asia in summer heats up and creates its own strong low-pressure system. This draws winds toward the continent from the warm Pacific and Indian oceans and as far away as Africa and Australia. In winter, the northeastern parts of the Asian continent become bitterly cold and generate a strong high-pressure cell over eastern Siberia. Winds blow out of the high and across the continent in the opposite direction of summer air flows. When a monsoon passes over the ocean, it picks up water vapor and brings rain to the land. If it blows from land, it is a dry wind. Depending on location, the same monsoon season may be either dry or wet. When onshore tropical monsoonal winds meet high mountains, rain in excess of 400 in (10,000 mm) can result, all of it falling in just a few months.

Precipitation Controls

Precipitation happens when the water vapor in saturated air condenses and millions of tiny water droplets coalesce to form larger drops that are too heavy to remain buoyant and fall to Earth. Most water vapor in the atmosphere comes from

Figure 4.6 Opposite sides of the Sierra Nevada clearly show the contrast in climate and vegetation on windward versus leeward slopes: (left) Giant sequoias grow on the wet western side of the range where orographic uplift of air coming off the Pacific Ocean creates frequent low clouds and high amounts of precipitation. (upper right) The rainshadow effect on the eastern, leeward side produces a desert in Owens Valley. (lower right) Extreme aridity occurs farther east and beyond yet another range of mountains in Death Valley, parts of which lie below sea level. *(Photos by author.)*

the evaporation of water from large bodies water, particularly the oceans. Some does come from the land, primarily from plants, but also from soils. Plants release the water vapor produced as a by-product of photosynthesis through the pores (stomata) in their leaves, a process called transpiration. The combination of transpiration from plants and evaporation from soil moisture is known as evapotranspiration. Climatologists consider potential evapotranspiration when determining water budgets and distinguishing humid from arid climate types. Potential evapotranspiration refers to the maximum amount of moisture that could enter the atmosphere under local environmental conditions were the soils to be saturated. In semiarid and arid climates, annual potential evapotranspiration is greater than total annual precipitation.

If an airmass is to release its moisture it must first become saturated with water vapor; its relative humidity must reach 100 percent. At that point, it is holding all the water vapor it can. Any excess will condense into liquid water. How much water vapor is required to saturate an airmass depends on air temperature. The warmer the air the more water vapor it can hold and thus the more is needed to reach the saturation point. To reduce the amount of water vapor that can be contained in particular parcel of air—whether saturated or unsaturated—and force condensation (and eventually precipitation), its temperature must be lowered. The temperature at which an airmass becomes saturated and condensation begins is called dew point. Dew point varies according to the original water vapor content of the air. If the air is near saturation, dew point is only slightly lower than current air temperature; if the air is dry, dew point may be many degrees lower.

The cooling that produces precipitation is achieved by forcing air to rise higher into the atmosphere. As air rises above sea level, it comes under less pressure and expands; that is, the gas molecules move farther apart. With greater distance between molecules, the number of collisions between molecules is reduced and the vibration of individual atoms is diminished. Since temperature is a measure of atomic vibrations, air temperature lowers. These processes that change air temperature by simply causing air to rise (or descend—which compresses the air under greater atmospheric pressure, increases the collisions and vibrations, and raises the temperature) are called adiabatic processes. No loss of heat is involved, only changes in air density. Adiabatic rates of cooling or warming are constant, although rates differ for saturated and unsaturated air. Condensation involves a release of latent heat energy caught up in water molecules during evaporation. Once released, the latent heat becomes sensible heat and adds warmth to the airmass, counteracting some of the cooling. Unsaturated or dry air changes temperature at the rate of 5.5° F per 1,000 ft (10° C per 1,000 m); saturated air cools more slowly at a rate averaging about 3.3° F per 1,000 ft (5° C per 1,000 m).

Rising air and adiabatic cooling are essential prerequisites for precipitation. Uplift of air occurs in several important ways, each with a geographic and sometimes a seasonal pattern. Air rises when it is heated, since warming forces it to expand. This is what happens in a hot air balloon when the burners are turned on. The expanded, lighter air rises through the denser, cooler air surrounding it. The process is called convective uplift and is a common phenomenon every afternoon in the tropics and on summer afternoons in the mid-latitudes. Strong updrafts can result in the formation of thunderheads, but even weak updrafts produced this way produce puffy cumulus clouds, the visible signs of condensation.

When, as part of the global atmospheric circulation pattern, airmasses that originate at different latitudes come together, the converging air has no option but to rise. In equatorial latitudes, warm air from the subtropical highs of the Northern and Southern hemispheres come together and rise, forming the ITCZ. The high amounts of rainfall associated with the wet tropical climate are largely produced by this convergent uplift. Since the ITCZ shifts its latitudinal position with the seasons, it is the chief rainmaker in the seasonal wet and dry tropical climate as well.

The Polar Fronts of both hemispheres are other places where airmasses from different source areas meet. Rather than airmasses of similar temperature and density converging, as is the case at the ITCZ, at the Polar Fronts, airmasses of different temperatures and hence different density come together. The warmer, less dense air from a subtropical high is forced to rise over colder, denser air emanating from a polar high. The warm air cools as it rises, and the water vapor held in it condenses to form first clouds and, if cooled enough, rain or snow, depending on the time of year. Such frontal uplift creates the cloudiness associated with the subpolar lows and the rapid alternation between sunny and overcast weather known in the mid-latitudes.

The final type of uplift is more localized and not symptomatic of particular latitudes. Orographic uplift occurs as air passes upslope on the windward sides of mountains. The associated adabatic cooling makes the side of a range facing oncoming winds (the windward side) wetter than the leeward sides, where downslope air flows cause air temperatures to rise adiabatically. The warmed air can now hold more water vapor than it contains, so it becomes unsaturated or dry air and tries to absorb more water through evaporation rather than release any through condensation and precipitation. This produces the rainshadow effect.

Climate

Climate refers to the general picture of average or expected weather conditions over the course of a year at a particular location, whereas weather refers to the minute-by-minute or day-by-day changes in atmospheric conditions. A weather report gives readings of current atmospheric or barometric pressure, wind, temperature, relative humidity, dew point, and precipitation and may identify trends (such as rising pressure) and predict conditions for the next 24 hours or several days. Descriptions of climate deal in monthly and annual means or totals and concentrate on two variables, temperature and precipitation.

Annual patterns of these two variables are most important and are easily represented graphically as climographs that simultaneously display average monthly temperatures and precipitation (see Figure 4.7). Each climate type has a fairly distinctive climograph (see Figure 4.8), although much variation exists among climate stations in the real world. Sample data for stations representative of the world's major climate types and biomes are available in the Appendix.

For temperature, the concern is with the range in temperature between the coldest and warmest months; the coldest and warmest monthly temperature averages; whether or not freezing temperatures are experienced and for how long; and the length of the growing season, which is defined as the time between the last killing frost of spring and the first killing frost of autumn.

Precipitation is described in terms of total annual precipitation and how that amount is distributed through the year. In colder climates, the amount of snowfall and the duration of the snow cover may be important. In some regions, significant amounts of precipitation—significant for the support of plant growth—fall each month. In other areas, distinct wet and dry seasons predictably occur each year. As long as sufficient moisture accumulates to support forests, the climates with year-round adequate precipitation and those with seasonally adequate precipitation are

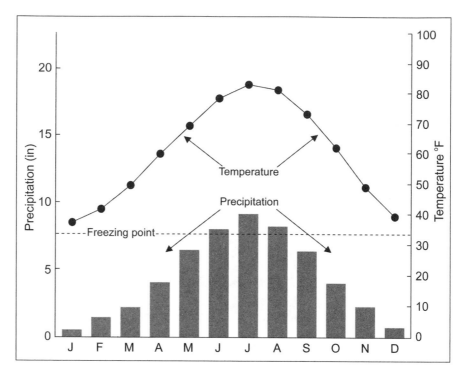

Figure 4.7 A generic climograph. *(Illustration by Jeff Dixon.)*

deemed humid. In areas where too little rainfall or snow cover usually occurs to support forests, grasslands and deserts prevail and the climates are considered semi-arid or arid, respectively.

Certain combinations of annual temperature patterns and annual precipitation patterns reoccur regularly over the Earth's surface. Each distinctive combination is a climate type. Since latitude is a major determinant of temperature patterns, humid climate classifications commonly reflect an association with specific latitudinal zones. The broad categories are tropical climates; warm temperate or subtropical climates; and cold temperate, snow, or continental climates. Within each of these climate types are major variants based on total precipitation amounts and, even more important, the seasonal rhythm of precipitation. The drier climate types, in contrast, are usually identified first according to total amounts of precipitation and secondary groups arise with the addition of temperature patterns.

Several climate classification systems have been developed. Each uses a somewhat different system of climate names and may utilize different parameters for drawing boundaries between regions characterized by different climate types. One of the most widely used classifications at introductory levels is that devised by the Russian-born German climatologist Wladimir Koeppen (1846–1940). Since he differentiated climate types and mapped them according to the vegetation's response

Plate I. Arctic Tundra Biome. (a) Tussocks of cottongrass (*Eriophorum angustifolium*) cover the low arctic tundra at Unalakleet, Alaska. *(Photo by author.)* (b) The snowy owl (*Nyctea scandiaca*) preys on lemmings and leaves the Arctic in years when lemming populations are low. *(Dennis Larson/USDA NRCS.)* (c) Caribou (*Rangifer tarandus*) and (d) muskoxen (*Ovibos moschatus*) are characteristic large herbivores with circumpolar distributions. *(Bob Stevens/USFWS; USFWS.)*

Plate II. Alpine Tundra Biomes. (left) Temperate Alpine Biome. (a) Arctic-alpine zone in the Beartooth Mountains, Montana. Many of the same plant species that are found in arctic tundra also occur here. *(Photo by author.)* (b) American pikas (*Ochotona princeps*) live in talus. *(John J. Mosesso/NBII.)* (c) Mountain goats (*Oreamnos americanus*) graze alpine pastures in the Rocky Mountain section of the biome in North America. *(Dave Grickson/USFWS.)*

(right) Tropical Alpine Biome. (d) Tall *Espeletias* or frailejones cover the slopes in the páramo of northern Ecuador. (e) Vicuñas (*Vicugna vicugna*) are one of two members of the camel family that can be found in the puna of Peru, Bolivia, and Chile. *(Photos by author.)*

Plate III. Boreal Forest Biome. (a) An evergreen needleleaf forest dominated by spruce and fir stretches across Canada. Lakes and muskeg are prominent parts of the landscape. *(Courtesy of Voyageurs National Park/NPS.)* (b) Mink *(Neovison vison)*, one of many members of the weasel family trapped for their furs. *(Dennis Larson/USDA NRCS.)* (c) Moose *(Alces alces)* feed in wetlands and early successional communities where deciduous shrubs provide browse. *(John and Karen Hollingsworth/USFWS.)* (d) The red squirrel *(Tamiasciurus hudsonicus)* feeds on pine nuts. *(Donna Dewhurst/USFWS.)* (e) The whiskeyjack or gray jay *(Perisoreus canadensis)* is common. *(John J. Mosesso/NBII.)* (f) The lynx *(Lynx canadensis)* is one of the larger predators and was another valuable furbearer. *(Erwin and Peggy Bauer/USFWS.)*

Plate IV. Temperature Broadleaf Deciduous Forest Biome. (a) Forests of the Blue Ridge Mountains in Virginia are typical of the biome in eastern North America. *(Photo by author.)* (b) The several layers of green plants include the herb layer on the forest floor, a shrub layer (here of rhododendron), a sapling layer, and a canopy layer. *(Photo by author.)* (c) The Black-throated Blue Warbler *(Dendroica caerulescens)* represents one of many Neotropical migrants that nest in the forest. *(S. Maslowski/USFWS.)* (d) The Blue Jay *(Cyanocitta cristata)* is a resident species. Its habit of storing acorns in the ground probably helped reforest large tracts of land when farmland was abandoned. *(Dave Menko/USFWS.)* (e) The sessile bellwort or wild oats *(Uvularia sessifolia)* is representative of the many delicate members of the herb layer. *(Photo by author.)* (f) Grey squirrels *(Sciurus caroliniensis)* are agile climbers and leapers and are common in the forest canopy. *(John J. Mosesso/NBII.)*

Plate V. Mediterranean Woodland and Scrub Biome. (a) Firebreaks cut across the highly flammable chaparral in the Santa Monica Mountains near Los Angeles. *(Photo by author.)* (b) The California Gnatcatcher *(Polioptila californica)* is an endemic and endangered species of the coastal sage section of the California chaparral. *(B. Moose Peterson/USFWS.)* (c) Chamise *(Adenostoma fasciculatum)* is perpetuated by frequent fires and resprouts from its roots. It is shaded out by taller plants if fires are prevented over long periods of time. *(Photo by author.)* (d) *Brodiaea californica*, a perennial forb of the chaparral, puts forth abundant blooms after a fire. *(Photo by author.)* (e) Yucca *(Yucca whipplei)* demonstrates both a rosette growthform that helps protect the renewal bud from fire and the ability to resprout after a burn. *(Photo by author).* (f) Protected from fire and browsing animals, an elfin forest of oaks has developed on Santa Cruz Island off California. *(Photo by author.)*

Plate VI. Temperate Grassland Biome. (a) The pronghorn *(Antilocapra americana)* is the sole member of a mammal family endemic to North America and not a true antelope. *(Gary Kramer/USDA NRCS.)* (b) Bison *(Bison bison)* graze short-grass prairie on the National Bison Range in Montana. *(Photo by author.)* (c) The Western Meadowlark *(Sturnella neglecta)* sings a different song than the nearly identical Eastern Meadowlark. *(Jeff Vanuga/USDA NRCS.)* (d) The Russian steppe in its yellow aspect. *(Photo by author.)* (e) The American badger *(Taxidea taxus)*, a member of the weasel family, preys on rodents and ground-nesting birds. *(John J. Mosesso/NBII.)* (f) The black-tailed prairie dog's *(Cynomys ludovicianus)* burrows provides habitat for other fossorial animals and is a major prey species for black-footed ferrets, hawks, snakes, and other predators. *(Gary M. Stoltz/USFWS.)* (g) The endangered Attwater's Prairie Chicken *(Tympanuchus cupido attwateri)* has elaborate courtship rituals, as do other subspecies on the American prairies. *(George Lavendowski/USFWS.)*

Plate VII. Tropical Rainforest Biome. Neotropical rainforests display the great biodiversity characteristic of the biome. (a) The howler monkey (*Aloutta palliata*) is arboreal like most animals of the rainforest and has a prehensile tail. *(© Jan Kooreman. Used with permission.)* (b) Green iguanas (*Iguana iguana*) seek sunlight in the forest canopy. *(© Jan Kooreman. Used with permission.)* (c) Jaguar (*Panthera onca*), the largest predator in Neotropical rainforest. *(© Phil Palmer/www.BirdHolidays.co. uk. Used with permission.)* (d) Squirrel monkey (*Saimiri oerstedii*). *(© Jan Kooreman. Used with permission.)* (e) Leaf cutter ants (*Atta* sp.) grow fungi in their underground nests. *(Photo by author.)* (f) Epiphytes and large-leaved climbers drape the trees in Brazil's Atlantic rainforest. *(Photo by author.)* (g) Green-headed Tanager (*Tangara seledon*), one of many tanagers inhabiting the lowland and montane rainforests of the Neotropics. *(© Phil Palmer/www.BirdHolidays.co.uk. Used with permission.)*

Plate VIII. Tropical Seasonal Forest Biome. (a) Caatinga in the rainy season and (b) in the dry season. (c) Tropical seasonal forest in Sinaloa, Mexico, looks much like tropical rainforest during the rainy season, but is simpler in structure. (d) Columnar cacti are members of the seasonal forest in Oaxaca, Mexico. *(Photos by author.)*

Neotropical mammals: (e) Baird's tapir (*Tapirus bairdii*). *(Photo by author.)* (f) Marmoset (*Callithrix* sp.). *(© Phil Palmer/www.BirdHolidays.co.uk. Used with permission.)* (g) Collared peccary (*Tayassu tajuca*). *(John J. Mosesso/NBII.)*

Plate IX. Tropical Savanna Biome. East African savannas are known for their high diversity of animals, especially large mammals. (a) Plains zebra (*Equus burchelli*), (b) termite (*Macrotermes* sp.) mound, (c) nests of Sociable Weavers (*Philetarius socius*), (d) Hippopotamus (*Hippopotomus amphibius*), (e) white or square-lipped rhinoceros (*Ceratotherium simum*), and (f) African elephant (*Loxodonta africana*). *(Photos by author.)*

Plate X. Tropical Savanna Biome. Neotropical savannas lack the large mammals so characteristic of Africa. (a) Cerrado in Brazil. *(Photo by author.)* (b) An upland palm savanna in Sinaloa, Mexico, probably the product of repeated burning and grazing of tropical seasonal forest. *(Photo by author.)* (c) Greater Rhea (*Rhea americana*). *(© Roy Slovenko. Used with permission.)* (d) Giant anteater (*Myrmecophaga tridactyla*). *(© Roy Slovenko. Used with permission.)* (e) Capybara (*Hydrochaeris hydrochaeris*), a semiaquatic rodent and the world's largest rodent. The Giant Cowbird (*Molothrus oryzivorus*) perched on its head picks flies off capybaras. *(© Phil Palmer/www.BirdHolidays.co.uk. Used with permission.)*

Plate XI. Warm Desert Biome. The three warm deserts of North America: (a) the Sonoran, (b) the Mohave, and (c) the Chihuahuan. (d) The Roadrunner (*Geococcyx californianus*) gains most of its fluids from its diet of lizards. Reptiles including (e) desert tortoises (*Gopherus agassizii*) and (f) rattlesnakes (*Crotalus* spp.) are pre-adapted to life in arid environments. (g) Creosotebush (*Larrea tridentata*) is the indicator plant of warm deserts in North America. (h) The desert bighorn sheep (*Ovis canadensis nelsoni*) must drink water every few days. (i) The fog desert in Baja California, Mexico, is characterized by boojum trees (*Fouquieria columnaris*) and considered a section of the Sonoran desert. *(Photos by author except (d) Gary Kramer/USFWS; (h) Lynn B. Starnes/USFWS.)*

Plate XII. Cold Desert Biome. The Great Basin Desert, with sagebrush as the indicator plant, is the cold desert of North America. Many of the cold desert animals are also inhabitants of the prairies or of neighboring forests. (a) Coyote (*Canis latrans*). *(John J.Mosesso/NBII.)* (b) The desert above the gorge of the Rio Grande River in New Mexico. *(Photo by author.)* (c) Black-tailed jackrabbit (*Lepus californicus*). *(George Harrison/USFWS.)* (d) Greater Sage-Grouse (*Centrocercus urophasianus*). *(Gary Kramer/USFWS.)* (e) Yellow-bellied marmot (*Marmota flaviventris*). *(John J. Mosesso/NBII.)* (f) Mule deer (*Odocoileus hemionus*). *(Gary Kramer/USDA NRCS.)* (g) Elk (*Cervus elaphus*). *(Lisa Zolly/NBII.)*

Plate XIII. Freshwater Biomes. The three freshwater biomes, lakes, streams, and wetlands, are often connected. (a) Most of the world's lakes were formed by glaciers as were these ponds in Maine. (b) Streams provide a variety of aquatic habitats as they undercut banks and deposit sands and gravels on the inside of meanders. Habitats vary along the length of the stream, also, as shaded headwater streams give way to broad high order mainstems. (c) Cranberry Glades, West Virginia, a true bog. *(Photos by author.)*

Mammals, birds, amphibians, fishes, and invertebrates inhabit freshwater biomes. (d) Muskrat (*Ondatra zibethicus*). *(Dave Menke / USFWS.)* (e) Black crappie (*Pomoxis nigromaculatus*). *(Photo by Eric Engbretson / USFWS.)* (f) Crayfish (family Astacidae). *(Photo by Eric Engbretson / USFWS.)* (g) Dragonfly (order Odonata). *(Steve Hillebrand / USFWS.)* (h) Green frog (*Rana clamitans*). *(John J. Mosesso / NBII.)* (i) Mallards (*Anas platyrhynchos*) with Canada Goose (*Branta canadensis*). *(Photo by author.)*

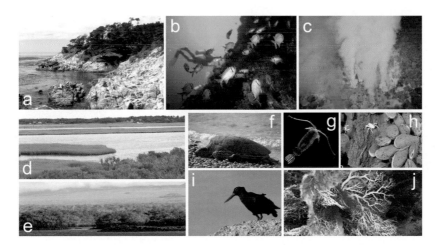

Plate XIV. Marine Biomes. The three marine habitats: (a) Coast Biome. The rocky coast habitats at Monterey, California, merge offshore with a subtidal kelp forest on the continental shelf. *(Photo by author.)* (b) Continental Shelf Biome. Coral reefs are biodiversity hot spots in clear, warm waters on tropical continental shelves. Here, squirrelfish (*Holocentrus adscensionis*) swim above corals in the Caribbean Sea. *(Dr. Anthony R. Picciolo / NOAA NODC.)* (c) Deep Sea Biome. Champagne Vent, a deep sea hydrothermal vent. *(NOAA / Ocean Explorer.)*

(d) Saltmarshes and (e) Mangroves are vital nursery areas for marine fish and shellfish and protect the shore from erosion and the highly productive shallow coastal waters from pollution. *(Photos by author.)* Animals of marine biomes represent almost all phyla. (f) Green sea turtle (*Chelonia mydas*). *(Ryan Hagerty / USFWS.)* (g) Copepod (*Valdiviella insignis*). *(R. Hopcroft, University of Alaska–Fairbanks and Census of Marine Zooplankton.)* (h) Vent crabs, shrimps, and mussels. *(NOAA / Ocean Explorer.)* (i) African Oystercatcher (*Haematopus moquini*). *(Photo by author.)* (j) Deep sea gorgonian corals (*Corallium* sp.) with a deep purple octocoral (*Trachythela* sp.), brittle stars, crinoids, and sponges. *(Mountains in the Sea Research Team; the IFE Crew; and NOAA.)*

Spodosol   Alfisol   Ultisol

Mollisol   Oxisol   Histosol

Plate XV. Soil profiles for some major soil orders. Major zonal soils are recognized by the distinct characteristics of their horizons. Spodosols are associated with the Boreal Forest Biome; Alfisols and Ultisols with the Temperature Broadleaf Deciduous Forest Biome. The Mollisols develop best in the Temperate Grassland Biome. Oxisols are characteristic soils of both the Tropical Rainforest Biome and the Tropical Seasonal Forest Biome. Histosols are the mucks and peats that form in wetland situations. *(Photos courtesy of USDA Natural Resources Conservation Service.)*

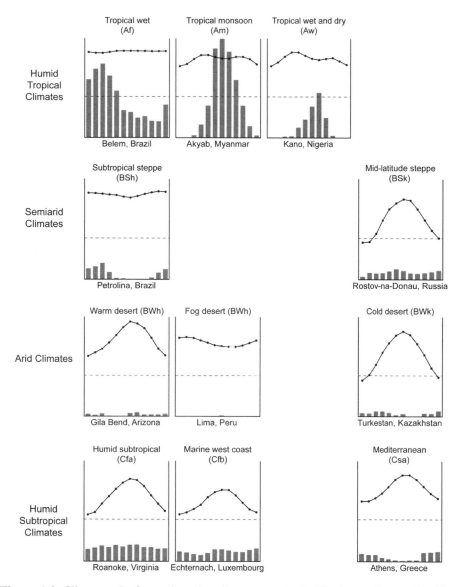

Figure 4.8 Climographs for each major climate type in the Koeppen climate classification. Compare with Table 4.1, and see the Appendix for actual precipitation and temperature data for each station. *(Illustration by Jeff Dixon.) (Continued)*

to temperature and precipitation patterns, the Koeppen system, with some later modifications (see Table 4.1), is especially useful in the study of biomes. Each climate type has its characteristic biome and vice versa (see Table 1.1). An overview of each major climate type follows. The shorthand symbols Koeppen employed are indicated as an aid to readers familiar with the system.

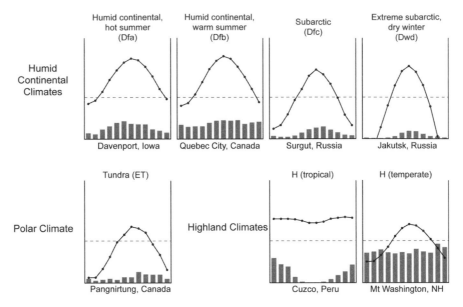

Figure 4.8 (*Continued*)

Humid Climates

Humid tropical climates. The essence of a humid tropical climate (an A climate in the Koeppen system) is year-round warm temperatures. Every month has an average temperature no lower than 64.4° F (18° C). Within this group are three major types, differing from each other in terms of how the rainfall is distributed during a normal year. Regions with Tropical Wet or Tropical Rainforest Climates (Af) receive at least 2.5 in (60 mm) of rain every month. Usually monthly totals are much higher than this, and total annual precipitation may be 100 in (2,500 mm) or more. Most parts of the world with this climate type occur in the equatorial latitudes. As one name for this climate type implies, these amounts of precipitation support tropical rainforests.

Another type of tropical climate that has enormous amounts of rainfall recorded each year is the Tropical Monsoon Climate (Am). Most areas with this climate type are influenced by the Asian monsoon and orographic uplift. The southwest or Kerala coast of India is one such place; another region encompasses parts of Bangladesh, Myanmar, and Malaysia. Other monsoonal regions occur in South America just north of the mouth of the Amazon River and in coastal West Africa in Guinea, Sierra Leone, Liberia, and southern Côte d'Ivoire. One to three months of the year are dry, each receiving less than 2.5 in (60 mm) of rain. However, their lack of precipitation is more than made up for by the months when the onshore monsoon brings heavy rains. In most cases, the natural vegetation is tropical rainforest. The record for total precipitation in one year is held by a weather station in the tropical monsoon climate. In 1860–1861, Cherripunji, India, received 1,042 in (26,470 mm) of rain during the summer monsoon.

Table 4.1 Major Koeppen Climate Types

NAME	SYMBOL	DESCRIPTION
HUMID TROPICAL CLIMATES	**A**	Total precipitation sufficient to support tropical forests and savannas; all months have mean temperatures of 64.4°F (18°C) or higher. Small annual temperature range.
Tropical Wet or Tropical Rainforest	Af	No dry season, total annual precipitation 60–100 in (1,520–2,540 mm), no month receives less than 2.5 in (60 mm).
Tropical Monsoon	Am	Short (1–3 months) winter dry season; total annual precipitation 100–200 in (2,540–5,080 mm); rains associated with monsoonal winds.
Tropical Wet and Dry or Tropical Savanna	Aw	Distinct wet and dry seasons, dry season 3–6 months long; total annual precipitation 35–70 in (900–1,800 mm). Rains associated with passage of ITCZ over region in summer.
DRY CLIMATES	**B**	Distinct seasons based on temperature differences; precipitation scarce all year, total annual precipitation less than 30 in (760 mm).
Semiarid or Subtropical Steppe	BSh	Hot summers, relatively mild winters; total annual precipitation 15–30 in (380–760 mm).
Semiarid or Mid-latitude Steppe	BSk	Continental temperature pattern: very hot summers and cold winters. Total annual precipitation 10–25 in (250–640 mm).
Arid or Subtropical or Warm Desert	BWh	Hot summers, relatively mild winters; rainfall scarce and sporadic; total annual precipitation less than 15 in (380 mm).
Arid or Mid-latitude or Cold Desert	BWk	Continental temperature pattern: very hot summers and cold winters. Total annual precipitation less than 10 in (250 mm).
HUMID SUBTROPICAL CLIMATES (temperate climates)	**C**	Distinct seasons based on temperature differences; mean temperature of warmest month above 50° F (10° C); sufficient precipitation to support forests and woodlands.
Humid Subtropical	Cfa	Hot summers—warmest month higher than 72° F (22° C), mild winters; total annual precipitation 40–60 in (1,000–1,500 mm); summer maximum, but year-round rainfall.
Marine West Coast	Cfb	Mild summers—warmest month less than 72° F (22° C), mild winters for the relatively high latitude; total annual precipitation 30–100 in (750–2,500 mm).

(Continued)

Table 4.1 (*Continued*)

Name	Symbol	Description
Mediterranean	Cs	Hot summers, mild winters; total annual precipitation 15–25 in (380–640 mm); summer dry season.
HUMID CONTINENTAL CLIMATES (snow climates)	**D**	Cold winters, summers 4–8 months long with mean monthly temperatures above 50° F (10° C), large annual temperature range; total annual precipitation sufficient to support forests, summer maximum. Weather highly variable from day to day.
Humid Continental, hot summer	Dfa	Hot summers [mean monthly temperatures above 72°F (22° C)], cold winters; total annual precipitation 20–40 in (500–1,000 mm); deep snowfall.
Humid Continental, warm summer	Dfb	Warm summers [mean monthly temperatures below 72° F (22° C)] with at least 4 months above 50 F (10 C); cold winters; total annual precipitation 20–40 in (500–1,000 mm); deep snowfall.
SUBARCTIC	**D**	Long, dark, and very cold winters, brief mild summers; extremely large annual temperature range; total annual precipitation low.
Subarctic	Dfc	Long, dark, and very cold winters, brief mild summers; extremely large annual temperature range; total annual precipitation 5–20 in (130–500 mm).
Subarctic, dry winter	Dwd	Extremely cold winters, mild summers; total annual precipitation low with very little snowfall in winter.
POLAR CLIMATES	**E**	Long, dark winters (24 hours of darkness during some months), very brief summers; annual temperature range great but smaller than in subarctic regions; no month has mean temperature above 50° F; total annual precipitation less than 10 in (250 mm). Most precipitation falls as snow.
Tundra Climate	ET	Long, dark winters; no month has mean temperature above 50° F, the limit of tree growth; total annual precipitation less than 10 in (250 mm).
Ice Cap	EF	Long, dark, extremely cold winters; cold summers, no month with mean above 32° F; total annual precipitation less than 5 in (130 mm). Strong katabatic winds typical in Antarctica.
HIGHLAND CLIMATES	**H**	Climate changes rapidly across short map distances as a result of elevational change.

Note: ITCZ = Intertropical Convergence Zone.

The third humid tropical climate type, the Tropical Wet and Dry Climate (Aw), has a prolonged dry season of three to six months. During each of these months the average precipitation falls below 2.5 in (60 mm). The rainiest months occur during the high sun period of summer, when the subsolar point and ITCZ are nearby. The dry season occurs in winter, when the rain-bearing ITCZ is in the opposite hemisphere. The broadleaf evergreen trees of tropical rainforests cannot tolerate such long periods of drought, although deciduous trees can. The more moisture-loving varieties of seasonal or dry tropical forest flourish in this climate type. However, the biome most associated with this climate type is the Tropical Savanna, in which tall grasses well adapted to drought form a continuous ground-cover beneath an open canopy of deciduous trees. So common is this vegetation type that its name is often used to designate the climate type, even though disturbance factors such as fire and grazing may actually be more important in determining the structure of the savanna than the long dry season.

Humid subtropical climates. Primarily located between 25° and 40° N and S, subtropical climates (C climates in the Koeppen system) have noticeable temperature differences between the summer and winter months. The winters are described as mild. Although snow may fall, the cover rarely lasts more than a few days. On the west coasts of continents "subtropical" climates extend to latitudes of 60° due to the moderating influence of warm ocean currents offshore. Within the subtropical latitudes, different variants of subtropical climate occur on the east coast versus the west coast of continents. The differences show up in both the total amounts and the annual distribution patterns of precipitation.

On the eastern sides of continents the true Humid Subtropical Climate (Cfa) is found. Precipitation is rather evenly distributed through the year with a peak during the summer months. Normally, no months are dry, and average annual precipitation is 40–60 in (1,000–1,500 mm). Summers are hot and humid. In winters, freezing temperatures are common, but they do not continue for months at a time. Nonetheless frozen soil prevents water uptake by plants, and winter is a nongrowing season requiring special adaptations among plants and animals if they are to survive. The typical vegetation is temperate broadleaf deciduous forest. On poor and sandy soils, evergreen pine forests may dominate the landscape.

On the western sides of continents in the subtropics, the climate is under the control of a subtropical high-pressure cell in summer and the Prevailing Westerlies and Polar Front in winter. Summers are dry. Nearly all of the 15–25 in (380–640 mm) of precipitation falls in the winter months. This is the only climate type in which precipitation is not associated with the warmest time of year. Since it is typical around the shores of the Mediterranean Sea it has become known as a Mediterranean Climate (Cs) regardless of where it occurs. The regions of the Earth with a mediterranean climate are widely separated and, except for the Mediterranean proper, are small in area. The vegetation, like the precipitation pattern, is unique to these regions and identifies the Mediterranean Woodland and Scrub Biome. Small leathery and often evergreen leaves characterize many of the shrubs and small trees and appear to be adapted to the short growing season that exists toward the end of winter, when temperatures become high enough and soil moisture remains great enough for plants to grow. However, fire and grazing pressures may be equally important in determining the nature of the vegetation in some parts of the biome.

A second, higher latitude subtropical climate on the west sides of continents is simply known as the Marine West Coast Climate (Cfb). Winters are especially warm for the latitude and summers are typically mild. Moisture is received all year, but there tends to be a winter peak. The constant onshore flow of moisture air in this zone of Prevailing Westerlies creates much cloudiness and much drizzle all year long. In the Americas, the Marine West Coast climate regions are very narrow, because they are sandwiched between the Pacific Ocean and major north-south oriented mountain ranges. Orographic uplift contributes 100 in (2,500 mm) or more of rain and snow a year. In contrast, in western Europe, the climate type extends far inland unhindered by mountains, since the higher ranges all have east-west orientations. Lesser amounts of precipitation, commonly between 30 and 50 in (750 and 1,250 mm), occur without mountains to force the westerly flow of air to rise. In North America the Marine West Coast Climate is associated with a temperate rainforest of giant conifers. The same is true for the region in South America, although the tree species are different. In Europe, a temperate broadleaf deciduous forest covered the Marine West Coast climate region before it was cleared for agriculture and industry.

Humid continental climates. These climates are found on both large landmasses of the Northern Hemisphere between the latitudes of 40° and 70° N. Continental climates (D in the Koeppen system) are marked by wide extremes in temperature between the coldest and warmest months. Four seasons—winter, spring, summer, and fall—are easily distinguished. Only four to eight months have average temperatures above 50° F (10° C), the temperature usually thought of as a threshold for plant growth. The winters are severe with temperatures well below freezing for months at a time and only a few brief hours of low-angle sunshine each day. Total annual precipitation averages 20–40 in (500–1,000 mm), becoming less the farther north the position. Most rain and snow is associated with storms emanating from the Polar Front, although convection and monsoonal processes account for some summer rains. The length of winter and the degree of warmth in summer define several variants of this climate type.

The namesake Humid Continental Climate (Dfa and Dfb) appears on the eastern sides of continents, usually between 40° and 55° N. Precipitation is well distributed throughout the year, although it peaks during the summer. Significant amounts fall as snow, which may cover the ground from November through March. In far northeastern Asia, the effect of the high-pressure cell that forms during winter over this region is felt. Bitterly cold air holds little moisture, so snowfall is limited. This gives rise to a dry winter variant (Dw) of the Humid Continental

Climate. The milder humid continental climates (Dfa) are associated with the Temperate Broadleaf Deciduous Forest Biome. Usually the dominant trees are different species than those found in parts of the biome under the influence of a humid subtropical climate. In the colder, more northern parts of the climate region, mixed temperate forests containing both needleleaf evergreen trees and broadleaf deciduous trees are present.

Stretching all the way across both Eurasia and North America north of the humid continental climate regions is a Subarctic Climate (Dfc, Dwd) region. Occurring between 50° and 70° N, it is distinguished by long and dark winters. North of the Arctic Circle, some days will have no sunshine at all. Six or seven months of the year, temperatures will average below freezing, and the coldest months have *average* temperatures below −36° F (−38° C). Lakes that freeze in early autumn will not thaw until May. The coldest temperatures reported outside Antarctica or the Greenland ice cap come from Siberia, where the record is −90° F (−68° C). Summer monthly average temperatures are commonly in the upper 50s or lower 60s, but daily maxima may reach the upper 90s. The greatest annual range in absolute temperatures (a daily summer maximum versus a daily winter minimum) also comes from this climate region. Verkhoyansk in eastern Siberia, Russia, holds the world record at 188° F (104° C). Its highest recorded summer temperature is more than 98° F (36.7° C), and its lowest recorded wintertime temperature is −90° F (−67.8° C). This is the domain of the great boreal forest or taiga with its mostly evergreen conifers. In the most extreme areas of Siberia, the typical forest gives way to stands of deciduous larches.

Dry Climates

Dry climates (the B climates in the Koeppen system) are widely distributed across the globe and influence life on nearly 30 percent of Earth's land area. A number of factors contribute to the development of climates characterized by low-precipitation and high-evaporation rates. The dominance of high atmospheric pressure accounts for aridity over vast areas in the subtropics. Location in the rainshadows of major mountain ranges is responsible for many desert and grassland regimes in the mid-latitudes. Great distance from a major source of atmospheric moisture, usually the sea, creates drylands in the interiors of several continents. The world's driest deserts, however, lie right next to the ocean on the west coasts of continents between roughly 15° and 30° latitude. Here the whorl of air circling out of subtropical high-pressure cells blows parallel to or away from the coast, so little saturated air comes onto the land. These latitudes are also the location of the cold eastern boundary currents. When warm, moist air crosses these cold ocean surfaces, condensation occurs at sea level and forms dense fogs. If the fog-laden air does move onto the hot land, the moisture quickly evaporates, and it requires uplift farther inland along the flanks of hills and mountains to wring the water out. Humidity may be high and temperatures cooler than expected for the latitude, but it rarely rains. Most of the water available to plants and animals in these coastal deserts comes from the fog, not precipitation. Fascinating adaptations allow life to exist in the severe environment of such coastal fog deserts.

Dryness in the Koeppen system is recognized by the occurrence of natural vegetation other than forests. The degree of aridity is not merely a matter of low amounts of precipitation through the year, but it is determined by potential evapotranspiration exceeding total annual precipitation. Potential evapotranspiration is that amount of moisture that could be taken into the atmosphere if water were not limited. It is a function of temperature, since hotter air is able to evaporate a lot more moisture than cooler air. Thus, 20 in (500 mm) of precipitation a year can support a boreal forest in subarctic temperatures, but only grasses in tropical regimes. Nevertheless, as a rule of thumb, dry climates can be said to be those that receive less than 25 in (630 mm) of precipitation a year. These climates are subdivided into semiarid climates receiving 10–25 in (250–630 mm) annually and arid climates receiving less than 10 in (250 mm) of precipitation a year. Each of these climate types are then further subdivided according to annual temperature patterns (see below). As a general pattern, semiarid climates surround arid ones and serve as a transition between truly dry and truly humid conditions.

In addition to being scarce, precipitation in dry climates is known for being unpredictable. The drier the climate the greater the fluctuation in total amounts from year to year and the spottier its distribution. Averages may mean little to creatures trying to survive under such variable conditions. They must adapt to prolonged drought and be able to respond quickly when rains finally do occur. Low total amounts also belie the fact that the rare storms are often intense and flash floods always a danger in dry stream beds.

Semiarid climates. The semiarid climates (BS in the Koeppen system) are associated with natural grasslands outside the tropics, and as such are frequently called steppe climates. (In German, the word for grassland is *steppe*, which gives its initial letter S to the code for this climate type). In the subtropical semiarid climate type (BSh: h for *heiss* or hot, in German), the summers are hot and winters mild. Snow is rare. These regions are distinguished from cold or mid-latitude steppes (BSk: k for *kalt* or cold, in German), which in winter experience long periods of below-freezing temperatures and may receive much of their moisture as snow. Indeed these cold steppe climates have a continental temperature pattern with its wide range of temperatures between summer and winter. The Temperate Grassland Biome is mostly associated with the cold variant of the semiarid climate type.

Arid climates. The arid or desert climates (BW in the Koeppen system: W comes from the German *Wüste*, the word for desert) are similarly divided into warm or subtropical (BWh) and cold or mid-latitude (BWk) types. The world's highest temperatures have been recorded in the warm deserts of the subtropics and not at the Equator, where frequent cloudiness blocks incoming solar radiation. The record is held by El Azizia, Libya, in the Sahara. The temperature reached 136° F (58° C) on September 13, 1922. Similarly the greatest single-day temperature range comes from Sahara, where a difference of 100° F (56° C) separated early morning and mid-afternoon temperatures one day in 1927 at In-Salah, Algeria. Lack of moisture in the atmosphere lets much incoming solar radiation reach the ground during daylight hours and much outgoing heat energy escape at night.

Even so, the In-Salah record is an extreme case; more often the daily temperature range in subtropical deserts is closer to 20–50° F (11–27° C). Summer daily highs may frequently exceed 100° F (38° C), and most animals must seek refuge in the shade of shrubs or underground to prevent hyperthermia and severe water loss.

The west coast fog desert is usually considered a variation of the subtropical desert climate (BWh), although very different conditions exist. With cold currents offshore, summer temperatures do not get very high. The low 70s are common. Humidity is high and daily temperate ranges are low. These areas are deserts because they get little or no rain. Precipitation averages of less than 5 in (125 mm) a year are common and some places—for example in Chile's Atacama Desert— have supposedly never recorded rainfall. Arica, Chile, has an annual average of 0.03 in (0.8 mm) of rain, the world record for low precipitation. Perhaps surprisingly, even these driest deserts are not without life, although they may appear to be by the casual observer. Seeds of annuals buried in the soil are sometimes dormant for decades awaiting a rain that will let them germinate. Rosette plant forms are able to capture the fog and force condensation on their leaves or stems. The water droplets then drain down to their roots. Some beetles by virtue of behavior and morphology also catch the fog and funnel water droplets to their mouths. For anyone looking for the oddities of nature, fog deserts are great places to explore.

Polar Climates

Polar climates (E in the Koeppen system) are those with no month of the year having an average temperature above 50° F (10° C). They could be considered a type of dry climate, since the cold airmasses that dominate bear little moisture and annual average precipitation is generally less than 10 in (250 mm); however, they are treated as a separate category. Two types of polar climate are distinguished. One, the Tundra Climate (ET) is associated with a terrestrial biome; the other, the Ice Cap Climate (EF) is not. Both are high-latitude climates found in the polar regions of the world. Months of winter darkness and summer days 24 hours long are characteristic of both.

The Tundra Climate has at least one month of the year with an average temperature above freezing. Despite a more poleward latitude, winter temperatures never get as extreme as in the subarctic. The brief growing season is long enough to support slow-growing lichens and mosses and ground-hugging sedges, perennial forbs, and dwarf shrubs. The treelessness of the Tundra Biome has been attributed to the permafrost—the permanently frozen subsoil—just below the surface.

The Ice Cap Climate is only found on Greenland and Antarctica. Monthly average temperatures remain at or below freezing all year long. Precipitation is less than 5 in (125 mm) a year. The extreme cold produces high pressure over the ice caps, and very strong winds blow as the dense cold air moves downslope. These so-called katabatic winds blow the season's snow around in blizzard-like conditions. Much of northern North America and parts of Eurasia must have been subjected to similar pressure and wind systems during the Pleistocene Epoch when great ice sheets covered the land. No vegetation develops in these regions (although algae may grow in the snow) and so no biome is associated with this climate type.

Mountain Climates

In mountainous areas climate changes very rapidly across short horizontal distances. Atmospheric pressure and temperature decrease as elevation increases. Temperatures drop at an average rate of 3.5° F for every increase in altitude of 1,000 ft (6.5°C per 1,000 m). Exposure (the direction slopes face) affects temperatures as well. In the Northern Hemisphere, south-facing slopes receive more intense solar energy than do north-facing slopes and hence temperatures will be warmer. With warmer temperatures come higher evaporation rates, so south-facing slopes tend also to be drier. Winds are upslope on mountains sides facing into the wind and downslope on the leeward side, causing differences in total precipitation brought about by orographic and rainshadow effects, respectively. Even on the windward sides, precipitation may be unevenly distributed, increasing significantly at elevations above that level at which dew point is usually reached. In the tropics, in particular, at the elevations where the moisture in the air flowing upslope condenses, a line of clouds form. In the Spanish-speaking Andean countries this zone is known as *ceja andina*, the eyebrow of the mountains, and cloud forest vegetation develops there.

Because of the spatial complexity of mountain climates and the inability to display each type on a page-size world map, a separate category of Highland Climates (H) was added by later scientists to the Koeppen system of climate classification. This category clumps together all the variations into a single category that can be mapped and distinguished from surrounding regional climate types. The vegetation of regions with a Highland Climate forms distinct altitudinal belts or Life Zones. In the mid-latitudes of the Northern Hemisphere, the zones mirror the changes in vegetation encountered with latitudinal changes. In other words, climbing a high mountain is comparable to traveling from the subtropics through the subarctic to the treeline and—if the mountain is high enough—to the arctic. One passes from grasslands or broadleaf deciduous forests to boreal-like needleleaf evergreen forests, to alpine tundra and even permanent ice and

···

Altitudinal Zones in Latin America

In Spanish-speaking parts of Central and South America, four distinct climatic and vegetation zones are recognized: *tierra caliente* ("hot" land) at elevations generally below 3,000 ft (1,000 m); *tierra templada* (temperate land), an intermediate climate zone at elevations between 3,000 and 9,000 ft (1,000 and 3,000 m) where temperature patterns are similar to springtime in the north; *tierra fria* ("cold" land) rising above 9,000 ft to roughly 12,000 ft (3,000–4,000 ft) where freezing temperatures are frequent at night and treeline occurs; and *tierra helada* (frozen land), where it is always cold and alpine glaciers and snow fields persist. Actual elevations for all altitudinal zones vary with latitude and exposure, but even on the Equator, snow-capped mountains exist, although their glaciers are shrinking rapidly.

···

snow. In the tropics, one can proceed from tropical rainforests to ice caps, but the plant species in between are derived from tropical plant families, and the vegetation zones do not resemble those of northern latitudes. Above treeline in the tropics, alpine plants (and animals) must withstand daily freeze and thaw conditions, unlike arctic and alpine tundra species that cope with longer-term, seasonal temperature changes. Tropical plants are exposed to intense insolation about 12 hours a day year-round, whereas arctic plants must face months of darkness and months with low-angle sunlight all or nearly all day long. It should not be surprising, therefore, that tropical alpine and temperate alpine biomes are two distinct entities.

Climate Change

Today's climate types and distribution patterns have not always existed and will not exist in the future. Climate is the product of a complex system involving the Earth's relationships with the sun, the atmosphere, the continents, the oceans, and the presence or absence of large areas of ice. A change to one part has reverberations throughout the system. A single cause can never be identified, because each variation triggers response in other components of the system that can push yet other elements to new equilibriums. Climate changes significant in the distributional and evolutionary patterns of life on the planet occurred in the distant geologic past and in the recent geological past (during the Quaternary). Now change is happening again.

The causes or "forcing agents" of climate change operate at different timescales. Some require millions of years for their effects to be felt; others elicit responses in the atmosphere in briefer time periods from as little as a year to as many as 10,000 years. Those that happen over long stretches of geologic time are generally related to changes in Earth's surface and not to changes in the amount of solar radiation produced by or received from the sun. These "nonradiative" forcing agents are largely related to plate tectonics (see below). The continents have shifted their latitudinal positions over time; they have coalesced into supercontinents and then fragmented again. The amount of the continental plates above sea level has varied. Mountain-building and crustal uplift has raised the elevation of the land and modified pressure and wind systems. Similarly, the size and position of ocean basins have changed. In periods of especially active sea floor–spreading, the expanding magma chambers beneath mid-oceanic ridges raised the sea floor and, with it, sea levels. Changes in the configuration of land and sea altered the direction of ocean currents that transfer excess heat from equatorial regions poleward.

Climate changes that occur over shorter periods of time are usually caused by changes in the amount and distribution of solar energy received at Earth's surface. These "radiational" forcing agents are related to variations in Earth's orbit around the sun, to fluctuations in solar output, to changes in the composition of the atmosphere (especially the amount of greenhouse gasses, including water vapor), to changes in surface albedo (that is, the reflectivity of the surface), or to volcanic activity and the infusion of particulates, aerosols, and gases into the atmosphere.

Climate change in the distant geologic past. One of the earliest episodes of climate change began about 530 million years ago (mya) in the early Paleocene Period (see Figure 4.9).

At this time most of the continental masses were situated south of 30° N, and the continental cores of Antarctica, South America, Africa, and Australia were joined as the supercontinent Gondwana. Mid-oceanic ridges were expanding after the breakup of an earlier supercontinent (Rodinia) and raising sea levels to an all-time high. Many continental areas were flooded by shallow seas, and the abundance of water created maritime climates. It is possible that increased volcanic activity associated with rifting and plate collisions spewed carbon dioxide (CO_2) into the atmosphere. Indications are that CO_2 levels were 10 times higher than today. Even so, by the Ordovician Period (440 mya) and again in the Carboniferous (305 mya) glaciers covered much of Gondwana.

The Mesozoic was again a time of change. By the beginning of the era, in the Triassic Period (220 mya), the great supercontinent of Pangea had formed, tying all the continents together in one part of Earth. The huge landmass was surrounded by a single world ocean, the Panthalassa Ocean, but coastlines and continental shelf areas were reduced greatly when the several pieces of continental crust became a single supercontinent. The interior of the giant landmass must have exhibited great continentality in temperature patterns and great aridity. Almost as soon as Pangea formed, it began to be wrenched apart by tectonic forces. The Tethys Sea circled the equatorial regions between Laurasia to the north and Gondwana, again separate, to the south (see Figure 5.5). Rising sea levels flooded large parts of Laurasia, including parts that are now western Europe, North Africa, and North America. The Cretaceous Period was a time of worldwide warmth.

By the early Cenozoic Era, tropical temperature conditions extended poleward to 35°–40° latitude in both hemispheres. The continents were drifting northward, however, and by the late Eocene and Oligocene periods, the global climate was markedly cooler. The contemporary continents that made up Gondwana had completely separated from each other, resulting most importantly in the opening of the Drake Passage between the southern tip of South America and the Antarctic Peninsula on Antarctica some 30–25 mya (see also Chapter 5). Through the passage flowed a cold, circumpolar ocean current that cut off any southward transfer of heat from the tropics. The build up of the Antarctic ice cap may have started between 40 and 25 mya. Another period of cooling in the early Miocene (10–15 mya) coincides in the geologic record with a time of lowered sea level, suggesting the presence of major ice sheets in the polar latitudes of both hemispheres. It also coincides with the uplift of the Tibetan Plateau and Himalayan Mountains, as the Indian subcontinent merged with and subducted under the Asian continental plate. It has been hypothesized that the Tibetan Plateau reached its current elevation—an average 16,400 ft (5,000 m) above sea level—about 8 mya. Widespread climate changes may have resulted. The origins of the Asian monsoon trace to this time. The Gobi and Mongolian deserts and steppes became drylands as a consequence

Era	Period	Age (Ma)	Epoch	Major Geologic Events	Major Biologic Events
Cenozoic	Quaternary	0.01 2	Holocene	Widespread glaciation (ice ages)	Humans become ecologically dominant Extinction of megafauna First modern humans
Cenozoic	Quaternary		Pleistocene	Widespread glaciation (ice ages)	
Cenozoic	Tertiary	5	Pliocene	Isthmus of Central America forms Himalayas rise as India collides with Asia	
Cenozoic	Tertiary	26	Miocene		Grasses become abundant Artiodactyls diversify and spread
Cenozoic	Tertiary	37	Oligocene		
Cenozoic	Tertiary	53	Eocene		First perissodactyls
Cenozoic	Tertiary	65	Palaeocene		
Mesozoic	Cretaceous	136		Laurasia and Gondwana break apart	Rise and spread of flowering plants Extinction of dinosaurs
Mesozoic	Jurassic	190			First birds
Mesozoic	Triassic	225		Laurasia separates from Gondwana	Age of dinosaurs begins First mammals appear
Palaeozoic	Permian	280 345		Pangea forms	Mass extinction Rise of reptiles
Palaeozoic	Carboniferous				First gymnosperms
Palaeozoic	Devonian	395			First amphibians Age of Fishes
Palaeozoic	Silurian	430			First vascular land plants
Palaeozoic	Ordovocian	500 570			First fish
Palaeozoic	Cambrian			Rodinia breaks apart	
Precambrian	Proterozoic	2300		Rodinia forms	First multi-celled organisms
Precambrian	Archean				First unicellular lifeforms

Figure 4.9 Geologic time chart. *(Illustration by Jeff Dixon.)*

of the rainshadow effect of the highlands. The dust lying exposed to the winds in these arid lands was carried into the atmosphere to be deposited near the great bend in the Hwang He as loess. Some suggest that the rise of the Tibetan Plateau and Himalayas was a major force in the onset of the Ice Ages of the Quaternary Period. Glaciation in the Northern Hemisphere traces to the late Cenozoic, toward the end of the Pliocene, only 3–2 mya.

The end of the Pliocene also saw the formation of the Central American isthmus connecting North America to South America. The warm equatorial current of the Atlantic Ocean was thereby diverted northward as the Gulf Stream and, in its continuation as the North Atlantic Drift, it transported tropical heat to the middle and high latitudes of northwest Europe. (Westerly winds off the North American continent are also implicated in the maritime warmth of western Europe.)

The climate changes of the distant geologic past and the changes in the Earth's surface that caused them had major ramifications for life. Mass extinctions of higher taxa of plant and animal life accompanied major episodes of environmental change and paved the way for the adaptive radiation of the survivors. Land links between now widely separated continents allowed for the exchange of terrestrial and freshwater species and helped to create the taxonomic patterns revealed among shared taxa in Zoogeographic Provinces and Floristic Kingdoms. Fracture and isolation of continents provided opportunities for separate evolution and the rise of endemic families and orders. The same processes, often in reverse, affected life in the sea.

Climate changes in the Pleistocene. The cooling in the Northern Hemisphere during the late Pliocene heralded the Pleistocene Epoch (2 million to 10,000 years ago) of the Quaternary Period, known for its repeated cycles of glacial period cooling and interglacial period warming. These cycles are attributed either directly or indirectly to three variations in Earth-sun relationships:

- The tilt of Earth's axis (obliquity) varies from 22° to 24.5° over a 41,000-year period. Currently, of course, it is about 23.5° from the vertical, which is why the tropics lie between 23.5° N and 23.5° S. Changes in the tilt alter the latitudinal distribution of solar radiation both geographically and temporally. A decreased tilt decreases the amount of isolation received at high latitudes during the summer months and could result in the accumulation of ice in polar regions.
- The Earth's axis wobbles. In other words, the orientation of the axis changes as it pivots around its central point because of the gravitational pull of the sun, moon, and other planets on Earth's equatorial bulge. The tips of the axis make a complete rotation in roughly 23,000 years. Now the northern tip points to Polaris, our current North Star, but at the opposite position in the wobble, the star Vega becomes the North Star.

  The orientation of the axis affects the dates of the equinoxes and solstices. The shift was noted by ancient Greek astronomers who tracked the equinoxes against the backdrop of the constellations. They referred to this phenomenon as the precession of the equinoxes, a term still used. Precession of the equinoxes can affect the intensity and contrasts of the seasons.

- The orbital path of Earth around the sun changes shape over time in response to the gravitational pull of Saturn and Jupiter. Always an ellipse with the sun at one focus, Earth's orbit changes from nearly circular to a maximum eccentricity at periods of about 96,000 years. (A combination of factors and cycles is involved to reach this average.) Eccentricity is a measure of how far an ellipse deviates from a true circle. The variation in Earth's orbit ranges between eccentricities of 0.005 and 0.058 (currently it is 0.017). The shape of the orbit determines the distance that the planet lies from the sun and the length of seasons. Increased eccentricity lengthens the seasons that occur on the part of the orbit closest to aphelion, the point at which Earth is most distant from the sun.

   Currently aphelion is reached on July 4, so in the Northern Hemisphere, spring and summer are longer than winter and autumn. In fact, summer lasts 4.66 days longer than does winter. When the orbit changes so that the land-dominated Northern Hemisphere experiences winter and autumn at aphelion, this could significantly cool its middle and high latitude.

A Serbian mathematician, Milutin Milankovitch, related the patterns of these three variations in Earth-sun relationships to the approximate 100,000-year cycle of glacials and interglacials during the Pleistocene. Support for what is now called the Milankovitch cycle comes from oxygen isotopes in marine oozes, but many scientists regard the orbital changes as only regulating or tempering cycles brought on by internal interactions and feedback loops within the complex global climate system and not actually causing the cycles of glaciation and climate warming. Changes in CO_2 levels in the atmosphere are often viewed as the main triggers in Pleistocene climate change.

Glacial periods during the Pleistocene developed relatively slowly and each lasted roughly 100,000 years. Change to interglacials seems to have been rapid, in as few as 40 years according to some current models. Each interglacial lasted, on average, 10,000 years. The last glacial episode—known in North America as the Wisconsinan, and in Europe as the Würm—began about 120,000 years ago. The continental ice sheets reached their maximum extent and thickness at 18,000 BP (years before present). Retreat of the ice was well under way by 14,000 BP, and rapid deglaciation occurred between 11,000 and 10,000 BP, marking the end of the Pleistocene Epoch and the beginning of the Holocene. Warming, however, was not a continuous process.

Right at the time of rapid warming and deglaciation at the end of the Pleistocene came a brief cold period, especially in northwestern Europe, which has become known as the Younger Dryas. It may have been initiated by the shrinking of the North American ice cap to the degree that the St. Lawrence River spillway opened and sent huge volumes of freshwater into the North Atlantic. The hypothesis is that ocean salinity in the northern waters would have been lowered so much that the North Atlantic Deep Water (NADW) Current would have stopped. With the deep water circulation cut off, warm Gulf Stream waters would no longer flow as far north, and higher latitudes in the Northern Hemisphere would experience rapid cooling until such time as the amount of meltwater became greatly reduced.

A mass extinction of the Earth's largest land mammals and flightless birds—the megafauna—coincided with the end of the Pleistocene. In North America, 35 genera of mammals weighing more than 100 lbs (50 kg) disappeared. Among them were several elephants (mastodons and mammoths), giant ground sloths, and 15 genera of ungulates, including native horses, tapirs, camels, large mountain goats, giant big-horned sheep and bison, and large four-horned pronghorns. The largest carnivores also vanished—sabertooth cats, American lion, American cheetah, dire wolf, cave bears, and short-faced bears. The open boreal forest that stretched across the continent in the late Pleistocene had resembled modern East African savannas in terms of faunal diversity.

Other continents also suffered megafaunal extinctions, though not at precisely the same time. South America's wave of extinction may have been a little later, but also included mastodons and giant ground sloths. Australia's giant kangaroos and wombats and flightless birds were lost considerably earlier (40,000–50,000 years ago). Even Africa, today the paragon of large mammal diversity, lost some 50 genera about 40,000 years ago.

Climate change and habitat change may have played a role in Pleistocene extinctions, but they do not seem to fully explain the space-time pattern. Hunting of the large herbivores by early humans may have been too much for populations of slow-reproducing large animals to withstand. The large carnivores would have lost their prey species and as a consequence their populations, too, collapsed. Lethal diseases may have spread through populations already stressed by environmental change and human hunting pressures. The reasons for the loss of so many large forms of life are still debated.

Regardless of the role of Pleistocene climate change in the extinction of the megafauna, it did cause major changes in the distributions of plants and animals and the species composition of communities. Pleistocene vegetation and animal communities were unlike those of today, which are a product of species migration since the beginning of the Holocene and of adaptations to contemporary environmental variables.

Holocene climate change. In the last 10,000 years, climate changes have occurred at time scales of 10 to 1,000 years. Among the causes postulated are volcanism, solar variability, and changes in ocean circulation. After the Younger Dryas, global warming proceeded, until around 6,000 BP, when the thermal maximum for the Holocene was reached. At that time, sometimes called the Hypsithermal or Climatic Optimum, only the Northern Hemisphere was affected and the higher temperatures occurred only in summer. (Today's global warming affects winter and nighttime temperatures.) Arctic treeline advanced closer to the shores of the Arctic Ocean than it does today, and some prairie plants were able to advance eastward through the Appalachians as evaporation rates rose and aridity increased in eastern North America.

Climate continued to fluctuate into the historic period. After an episode of post-hypsithermal cooling, a European Medieval Optimum occurred from 1100–1300 AD,

••

Pleistocene Re-Wilding

Most of North America's largest animals and the species that depended on them had disappeared by 13,000 years ago. With their loss, predator-prey relationships changed; dispersers of large seeds were gone; evolutionary forces were relaxed. (For example, the pronghorn may be the world's fastest land mammal because when it evolved American cheetahs were around to pursue them.) A truly wild America became domesticated, as humans became the ecological dominants, altering landscapes, eliminating some of the native species that had survived the Pleistocene extinctions, and introducing exotic species to new habitats developing at human campsites and settlements. People created a landscape that functioned either to benefit human beings or in response to human activities.

Some scientists now argue that twenty-first-century conservation should pursue a goal of bringing parts of the continent back to a wild state by restoring the ecological and evolutionary processes that were in place before any people arrived. Since the largest animals—mammoths, horses, camels, American cheetah and lion, giant land tortoises, and many more—are extinct, close relatives or proxies such as Asian elephants, feral burros, guanacos, African cheetahs, African lions, and the Bolson tortoise would necessarily have to replace them. Large tracts of land in the American Southwest and on the Great Plains could be managed like the highly successful game parks in South Africa with free-roaming animals, functioning ecosystems, and thousands of paying visitors a year.

Proponents of re-wilding advocate restoring species interactions not just species. They know it will require careful planning and experimentation to work not only ecologically but also economically and politically. The standard for what is "natural," they say, is the environment of the Pleistocene before humans arrived in North America and not the conditions just prior to the arrival of Europeans in 1492.

••

allowing the colonization of Greenland and Iceland by the Vikings. This was followed by the Little Ice Age, which brought several cold periods separated by warmer times between 1450 (and perhaps earlier) and 1850. Rivers froze over in winter in western Europe and in Virginia. Ice skating became a popular sport in the Netherlands and elsewhere. Famine gripped areas where agriculture failed. Greenland was abandoned.

Earth once again is experiencing a global change of climate, usually referred to as global warming. Since about 1850, temperatures have been trending upward, with occasional setbacks related to major volcanic eruptions that spewed dust and aerosols high into the air. The warming trend coincides with the increase in atmospheric CO_2 since the beginning of the Industrial Revolution, but cause and effect relationships and hence predictions for future change are not straightforward, since the atmosphere-ocean-ice-land climate system introduces many potential responses and feedback loops. Nor is CO_2 the only greenhouse gas acting as a possible forcing agent.

Since 1850, the concentration of CO_2 in the atmosphere has increased from 290 parts per million (ppm) to 359 ppm, an increase of 27 percent. (During interglacials of the Pleistocene, CO_2 levels rose from 200 ppm to 280 ppm, confirming a relationship between air temperature and CO_2 concentrations.) Much of this comes from natural sources such as forest fires, releases from ocean waters, and the respiration of animals. The rest is generated by human activities, including the burning of fossil fuels, cement production, and clearing of forests.

Oceans and forests absorb CO_2 and store it in biomass. CO_2 dissolves in the surface layer of the ocean, where it occurs as dissolved gas and as bicarbonate (HCO_3^{-1}) and carbonate (CO_3^{-2}) ions. The colder the water, the more CO_2 it can dissolve. CO_2 is also removed from the atmosphere by surface-dwelling phytoplankters for photosynthesis. Many marine invertebrates use dissolved forms in the production of their calcium carbonate shells, freeing up space so to speak for more to be dissolved in the water when they die and their shells settle to the bottom to become oozes and sedimentary rocks, long-term storehouses of CO_2. It has long been believed that the oceans would take up excess atmospheric CO_2 and buffer temperature increases. Evidence is accumulating that while dissolved CO_2 is increasing the world's oceans, it is setting the stage for acidification of the waters. Slight numerical changes in the water's pH may have major repercussions among animals dependent on calcium carbonate shells and pose severe threats to the biodiversity of the sea.

Trees take up CO_2 from the air and incorporate it into woody tissues that endure in some cases for centuries. So, too, do grassland soils with their abundance of living and dead roots. When forests are cleared and burned, the organic compounds are reduced to inorganic constituents and CO_2 is released to the atmosphere. When grasslands are plowed up, the decay of roots is hastened and CO_2 is produced as a by-product. Land use changes associated with modern agriculture and urbanization have thus released a lot of CO_2 into the atmosphere. Nearly 3.5 times as much is added by the combustion of fossil fuels in power plants and motor vehicles.

Methane (CH_4) is another powerful greenhouse gases. Its concentration in the atmosphere has increased nearly 130 percent in recent times. Wetlands, termites, and oceans release methane gas. So do the digestive processes of ruminants such as cattle, rice paddies, landfills, and the burning of biomass.

Nitrous oxide (N_2O) is long-lived in the atmosphere and enters primarily as a result of biological activities, especially in tropical soils and in the oceans. Manmade sources include cultivated soils, biomass burning, and industry. Since the beginning of the Industrial Revolution, N_2O concentrations in the atmosphere have increased about 13 percent.

Another group of greenhouse gases, the halocarbons, are entirely manmade. These are compounds of carbon and chlorine, fluorine, and bromine. Most famous are the chlorofluorocarbons (CFCs) implicated in the destruction of Earth's protective stratospheric ozone layer. Ozone is also a greenhouse gas, and its loss counteracts the warming effect of the long-lived halocarbons but allows cell-damaging ultraviolet radiation to penetrate the atmosphere and reach Earth's surface.

Predictions of future change are necessarily based on modeling experiments. Since all the internal variables influencing climate cannot be accounted for in a mathematical equation, those that repeated observations and known relationships suggest are most important are fed into the models and tested. Global Climate Models (GCMs) estimate that if CO_2 alone is responsible for temperature changes, a doubling of CO_2 will create a temperature rise of 2° F (1.2° C), but the countless feedback loops mean the relationships are nonlinear and understanding the total system is difficult. Actual temperature changes may be lower or much higher than predicted by simple simulations. Most GCMs forecast major negative impacts on life on the Earth, including the planet's ever-expanding human populations. And most predict rapid changes, probably too rapid for most plant and animals to adjust their distribution areas or gene pools to adapt and survive. The temperature gradient between the Equator and the poles will diminish as polar regions experience greater warming than tropical areas. Existing climate types in the Arctic and in tropical highlands may disappear by 2100. Novel patterns of temperature and precipitation may arise in the tropics and subtropics, including such biodiversity hotspots as the Amazon rainforest and Andean Cordillera. One recent model resulting from a collaboration among researchers at the University of Wisconsin-Madison and the University of Wyoming predicts the disappearance of nearly half of terrestrial climates and the appearance of new climate types on as much as 39 percent of Earth's total land area.

Past climate changes: how do we know? Obviously no one was keeping records of climate conditions in the Mesozoic or Pleistocene, and scientific instruments recording weather events have become commonplace only relatively recently. So much of what we know about past climates, the knowledge upon which we build to predict future changes, comes from other sources. Each type of record is best suited for particular geologic or historic periods and particular time scales.

For the distant geologic past, one must rely on the rocks. The different kinds of sedimentary rocks are deposited under different depositional environments. For example, evaporites such as gypsum and salt accumulated during dry periods; limestones formed of ancient coral reefs indicate warm sea surface temperatures; and coal is the product of warm, humid conditions. The particle sizes and thicknesses of beds composed of sediments washed into the sea from land reveal terrestrial weathering and erosion processes that accelerate during warm, wet periods and slow down during cold, dry periods. In organic calcareous oozes on the sea floor the ratios between different isotopes of oxygen (O^{16} and O^{18}) incorporated into calcium carbonate indicate ancient ocean temperatures.

Oxygen isotopes were important in linking Milankovitch theory to climate change in the Quaternary. Other records of change closer to us in time than the

Paleozoic or Mesozoic are held in a variety of living and nonliving materials. Ice cores taken from polar ice caps and high mountain glaciers preserve not only oxygen isotopes but also dissolved particulates and atmospheric gases trapped in air bubbles. It is actually possible to use modern instruments to read the CO_2 content of the Ice Age atmosphere as well as the preindustrial atmosphere of only 400 years ago. The thickness of annual bands of ice indicate seasons and how much snow fell each year.

In seasonal environments, especially in the mid-latitudes, trees produce annual growth rings that are wider in good years and narrower in drought years. Samples of annual rings from dead and living trees can be aligned beside each other and cross-matched to stretch the record of climate change back hundreds and even thousands of years. Bristlecone pines in the White Mountains of California have yielded a record going back 8,500 years. An even longer dendrochronology has been produced from oaks in Germany.

Radiocarbon dating has been a powerful tool in reconstructing the environments of the Pleistocene. The technique is based on the rate of decay of an isotope of carbon. The common form of carbon is C^{12}, a stable atom with a molecular weight of 12 and a nucleus composed of six protons and six neutrons. A much rarer form is C^{14}, which has six protons and eight neutrons. C^{14} forms in the atmosphere when cosmic rays bombard a nitrogen atom, destroying one of its seven protons and changing it into an eighth neutron. The C^{14} atom is unstable and decays radioactively, reverting back to a nitrogen atom. Both C^{12} and C^{14} are incorporated into carbon dioxide molecules in the atmosphere, with C^{12} occurring a trillion times as often as C^{14}.

The rate at which any radioactive material decays is measured in terms of it half-life, the time it takes one-half of the nuclei in a sample to assume a stable form. (Half-life actually indicates a probability of decay: each atom in the sample has a 50 percent chance of decaying during the time interval. Fifty percent of the remaining radioactive nuclei then have a 50 percent chance of decaying during the next half-life period, and so forth.) Radioactive C^{14} has a half-life of 5,730 years, which allows its decay to be traced back 40,000–50,000 years across the time span of the middle and late Pleistocene.

Living plants and animals take up and incorporate carbon from the atmosphere into their tissues. The proportion of C^{12} and C^{14} assimilated will reflect the trillion-to-one ratio found in the air. Once the organism dies, however, no more C^{14} is assimilated and that which is already part of organic compounds begins to decay at a constant rate, and the proportion of C^{12} to C^{14} changes over time.

Dead organic material—plant matter, bone, tissue, charcoal, peat, or humus— can be analyzed in an accelerator mass spectrometer. This instrument separates C^{14} from C^{12} and measures the current $C^{12}:C^{14}$ ratio. From this it can be determined when the organism from which the sample derives was alive and a date can be assigned to the material.

Radiocarbon dates are given BP (or "before present"), but present means 1950. Standard radiocarbon dates are written with plus or minus two standard deviations, for example: 11,400±400 BP. This is a statistical notation that indicates a 95 percent certainty that the actual date lies within the stated range.

Other methods for dating Pleistocene and Holocene materials include potassium-argon ratios, thermoluminescence, and dendrochronology.

It is the late Pleistocene-Holocene changes that most inform climate modeling and predictions of the consequences of ongoing global climate change. A variety of scientific disciplines are engaged in interpreting Pleistocene environments. Geomorphologists study glacial landforms such as moraines and periglacial features such as patterned ground, or old lake beds with their varves or sediment deposits built up season after season and playas with ancient shorelines high above them. They can track glacial retreats and advances on local and continental scales and verify cycles of wet and dry climates in areas not directly affected by the ice. The fossil record of mammals, birds, and herps (reptiles and amphibians) reveals what occurred where and when. Comparing the habitats that closely related animals occupy today, paleontologists can make at least preliminary statements about the climatic conditions that existed at the time and places now-extinct and still-extant species roamed on Earth during an Ice Age. One of the most valuable and precise tools, however, is the pollen record. Pollen is the male gamete of flowering plants, and it is resistant to weathering and decay. Preserved largely in lake bed sediments, but also in dry lakes and river beds, it is recovered by coring or sampling at various depths. Palynologists can identify pollen grains (see Figure 4.10) to genus level and sometimes species level and read a record of vegetation change at a particular site.

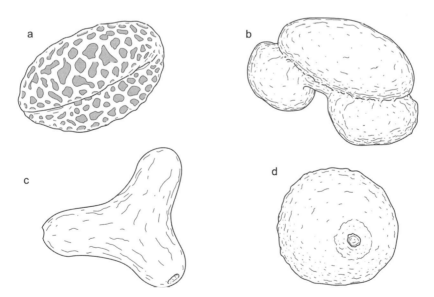

Figure 4.10 Pollen grains. Slow to decay, especially in lake bed sediments, pollen preserves a record of the vegetation that once grew near a collection site and how it changed through time. (a) A lily, (b) a pine, (c) a member of the evening primrose family, and (d) a grass. *(Illustration by Jeff Dixon.)*

Macrofossils of twigs, seeds, and needles may be present, as will charcoal and ash from fires. The organic matter can be radiocarbon dated. The results are most reliably interpreted as presence and absence data for plants, but from them inferences about vegetation change can be made that suggest climate change or changes in a disturbance regime that may be related to climate change.

Another record of Pleistocene vegetation change is held in fossil packrat middens in the drylands of North America. Packrats—or woodrats (*Neotoma* spp.)—deposit leaves, seeds, twigs, bones, and whatever else interests them in dry caves and crevices. Pollen, too, gets carried into the cave on plant material. The middens or garbage piles accumulate year after year and can become several feet deep. The rodents often urinate on their piles. Desert-dwellers needing to conserve body moisture, packrats produce a viscous urine that cements the debris together. As the urine dries and crystallizes, it forms an amber-colored coating known as amberat—and a distinctive odor. Under arid conditions, the material in the midden mummifies and is preserved for thousands of years (see Figure 4.11). Since packrats

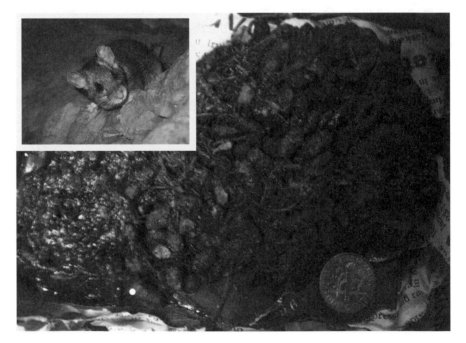

Figure 4.11 One source of information on Pleistocene vegetation and climate in the American Southwest has been fossil middens of packrats such as the brushy-tailed woodrat shown in the inset. *(National Park Service photo.)* The midden photographed was collected by Paul Leskinen at an elevation of 3,000 ft (1,000 m) in the Newbury Mountains of Nevada and was radiocarbon-dated at 13,380±BP, near the end of the last glaciation. Among the fecal pellets twigs and needles of juniper and pinyon pine are visible to the naked eye. Today these plants grow 1,000 ft (300 m) higher in the mountains. *(Photo by author.)*

generally collect materials only within a radius of 100–300 ft (30–100 m) of their homes, the middens offer a detailed record of local vegetation. Deposits less than 50,000 years old can be radiocarbon dated.

In rare instances fossil dung or coprolites of herbivores has been preserved in dry caves in the American Southwest. Since many herbivores, especially nonruminants, pass large masses of undigested plant matter through their digestive tracts, the boluses of dung can be analyzed relatively easily by the naked eye to reveal which plants the animals were eating. The microscopic techniques developed in the 1970s to analyze the grasses consumed by ruminants aid in the identification of grasses and other partially digested plant materials. The first dung deposit to be studied came from Rampart Cave in the Grand Canyon, where a vast accumulation of Shasta ground sloth (*Nothrotheriops shastensis*) boluses were discovered in the 1920s. Since then at least 30 other sites have yielded information on the diets (and hence the plant cover of the past) of other now-extinct large mammals, including Harrington's mountain goat, mammoth, and shrub oxen.

Climate change in the historic period before the development of modern scientific data collection methods is documented in written records that may be considered anecdotal. Farmers have always been vitally interested in weather events that

··

Phenological Changes Observed in Recent Times

Studies comparing the timing of spring events early in the twentieth century with what was happening near the beginning of the twenty-first century indicate that the growing season is beginning earlier and earlier in the mid-latitudes of the Northern Hemisphere. The calls of mating frogs began 10–13 days earlier in upstate New York in the 1990s than they did during the decade of 1900–1910. This was associated with a 1.8°–4° F (1°–2.3° C) rise in early spring temperatures. In the United Kingdom, 20 bird species laid eggs an average of 8.8 days earlier in 1995 than in 1971. A similar pattern was noted among Tree Swallows in New York state. One California study reported 17 of 23 butterflies were flying an average of 24 days earlier today than 30 years ago. Spring green-up in the eastern United States north of Virginia has occurred on average eight hours earlier each year since 1982.

These phenological changes all correlate with temperature changes, although other factors may also be involved. The response is strongest in the urban heat islands of cities, where springtime events may begin two to four days sooner than in nearby rural areas.

Many species have evolved life histories synchronized to those of other species. The threat from phenological changes in one species is that another species it depends on may not be changing its timing in a coordinated fashion. Predator-prey relationships can be disrupted, as can pollinator-plant associations. Returning Neotropical migrant birds in the temperate broadleaf deciduous forests of the Appalachian Mountains time their arrival to the budding of oaks, because this is when insects hatch or emerge and food is plentiful. Trees are leafing out earlier. It is still unclear whether insects and birds are changing their schedules.

··

influence planting and harvest schedules. They and gardeners and other plant enthusiasts have long kept records of flowering times, length of growing season, or the dates when different insects emerge. Especially people living in latitudes that experience long winters look for the first crocus to bloom, or the first robin to arrive in spring, or the first signs of green-up in the forest. Personal journals, old photographs, and newspaper reports all document these events. More systematic reporting is encouraged today, but scientists need these old sources to determine how present conditions differ from those of a century or more ago. The modern science of phenology studies the timing of plant and animal responses to seasonal climate changes. Satellite imagery and other remote-sensing techniques may be employed, but reliance is still heavy on on-the-ground observations by nonscientists. The National Phenology Network was established at the U.S. Geological Survey in 2007 to monitor changes in life's rhythms that are occurring in response to current global warming. Similar networks exist in other countries.

Plate Tectonics

It is well established that continents and ocean basins have changed their sizes, shapes, and geographic positions through geologic time and that they continue to do so. The likelihood that continents had once been attached to each other and then separated and moved to their present positions was proposed as early as the late-sixteenth century by the Belgian geographer Abraham Ortelius, better known as the cartographer who produced the first world atlas. A theory that both described and explained the phenomenon, however, was not put forth until 1915, when the German earth scientist Alfred Wegener (1880–1930) presented his hypothesis of continental drift. Wegener pursued multiple academic disciplines and brought together evidence from various sources to support his idea. As others had been, he was particularly intrigued by the "fit" between the coastlines of eastern South America and western Africa. Similar fossils and geology on both sides of the Atlantic Ocean were even stronger indications of a former connection between the two landmasses than map patterns. Wegener believed that 250 mya all the major continents were together as one supercontinent he named Pangea. Other scientists and new data collection techniques continued to build support for Wegener's theory, but it was not until the 1960s that it gained widespread acceptance. A major impetus came from American oceanographer Harry Hess (1906–1969), who identified and explained the phenomenon of sea-floor spreading and provided a mechanism by which the continents could be pushed around the globe. Today, the modern theory is embraced as plate tectonics and few question it.

Earth's outer layer, the lithosphere, is fractured into pieces of various sizes called plates (see Figure 4.12). Two different materials make up the upper-layer plates, oceanic crust and continental crust.

Oceanic crust is made continually as molten magma rises from the Earth's interior along the mid-oceanic ridges. New material breaks through the crust and

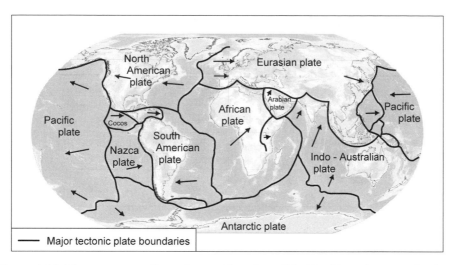

Figure 4.12 The current configuration and location of the Earth's major tectonic plates and the directions in which they are moving relative to each other. *(Map by Bernd Kuennecke.)*

pushes older cooled and consolidated material aside, widening the sea floor in the process. The oldest oceanic crust is drawn back down into the Earth or subducted at plate boundaries where two plates are colliding. The location of descent is often marked by oceanic trenches. The average length of time that oceanic crust remains at the surface is about 200 million years. Continental crust on the other hand is not dense enough to be subducted and is more or less permanent, although it is reworked by weathering, erosion, deposition, and metamorphism. The oldest known rocks exposed on the continents are 4 billion years old, but most land surfaces have much younger sedimentary or volcanic bedrock covering the ancient continental cores.

Plates with both oceanic and continental crust move, pulled along on convection currents in the molten mantle materials below the lithosphere. Plates diverge or move away from each other along those margins where magma wells up to the surface. Plates collide or converge with other plates or move past them as they go. Continental crust that cannot be subducted folds and breaks and is uplifted into mountain ranges along the boundaries between two converging plates. The movement is slow, but happens in jerks and is felt as earthquakes on the land. Sometimes the sudden movement occurs beneath the sea and generates tsunamis. Volcanic activity, including the construction of huge shield volcanoes, such as the Hawaiian Islands, or stratovolcanoes, such as those of the Cascades in North America or the ice-covered cones in the Andes, is another consequence of subduction. The movement of the plates, together with the processes that deform Earth's crust and create major landforms, are known as plate tectonics.

In the study of terrestrial biomes the importance of plate tectonics is in the latitudinal rearrangement of the continents that both created and ruptured links between landmasses and that changed climates in the geologic past. The rise of great mountain ranges such as the Himalayas, Andes, Alps, Atlas Mountains, Rockies, and Sierra Nevada also changed global and regional climates and opened corridors for migration or created barriers to the movement of plants and animals. As a consequence, new adaptations and sometimes species evolved, and other forms went extinct. The current vegetation and community structures as well as species compositions of all expressions of the world's biomes are thus products of the geologic past.

The details of plate tectonics are beyond the scope of this book. The interested reader is directed to books on earth science, geology, and physical geography for fuller descriptions and explanations. More information is also presented in the next chapter in which the consequences of plate tectonics on marine biomes are addressed. The basic pattern of continental drift is as follows. Before 450 mya five major continents probably existed, products of the break up of a former single landmass named Rodinia that existed some 900 mya. During the Paleozoic Era, these five continents came together to form Pangea. For perhaps 200 million years thereafter, only one continent and one ocean existed on Earth. Then early in the Triassic Period of the Mesozoic Era this supercontinent, too, began to split apart. Some scientists suggest that the huge landmass acted as an insulating cap on the lithosphere and allowed heat from the Earth's interior to build up beneath it. The mantle also would have been warmed, and convection currents would have been set into motion. Upwelling magma pushed up the crust and fractured it, allowing fresh lava to spill out on the surface and begin the process of sea-floor spreading. While the exact process remains to be discovered, it is known that Pangea first broke into two separate continents, a northern Laurasia and a southern Gondwana or Gondwanaland, as it is sometimes called. Each of these still-large continental masses later split apart. Laurasia separated into North America, Greenland, and most of Asia north of the Himalayas. Gondwana separated into South America, Antarctica, Africa, Australia, India, and Madagascar. The ancient ties of the southern continents are especially evident in shared plants and animals such as the southern beeches (genus *Nothofagus*), Southern Hemisphere pines (family Araucariaceae) (see Figure 4.13), and the Neotropical and Australian marsupials.

Gondwana itself may have been centered near the South Pole. The movement of its parts away from each other translates into a northward movement into tropical latitudes. Northern continents, too, pushed northward, reaching the mid-latitudes and high latitudes that they dominate today only a few million years ago. Smaller pieces of continental crust, sometimes called microcontinents, also broke off and later merged with the landmasses we know today. One of the larger microcontinents became the Indian subcontinent, which rather rapidly moved north across the Equator to collide with Eurasia and force the uplift of the Tibetan Plateau perhaps 40 mya and the rise of the Himalayan Mountains around 8–10 mya.

Figure 4.13 An araucaria forest near Campos do Jordão, Brazil. These tall coniferous trees are Gondwanan relicts and were in existence before the breakup of Pangea. Close relatives are found in Australia and New Caledonia. *(Photo by author.)*

Plate tectonics continues today, although some indications are that the process may be slowing. The Atlantic Ocean, which only came into being with the break up of Pangea some 250 mya, is widening at a rate of about 2 in (5 cm) a year. At the East Pacific Rise, sea-floor spreading is faster and has been measured at as much as 5 in (13 cm) a year. Recent reports hypothesize that the Pacific will disappear in about 350 million years when Asia, Australia, and the Americas converge and continental drift will then halt—at least for while.

Further Readings

Consult any physical geography or world regional geography textbook for more information on latitude and longitude, Earth-sun relationships, and climate.

Internet Sources

Atmosphere, Climate, and Environment Information Programme. 2002–2004. "Global Climate Change Student Guide." Manchester Metropolitan University. http://www.ace.mmu.ac.uk/resources/gcc/contents.html. An excellent compendium of information on climate controls and climate changes.

Canty and Associates. 2008. Weatherbase. http://www.weatherbase.com. Climate data for many stations throughout the world.

Cole, Kenneth. 2002. "Packrat Middens." In *Canyons, Cultures, and Environmental Change: An Introduction to the Land-Use History of the Colorado Plateau*, ed. John D. Grahame and Thomas D. Sisk. http://cpluhna.nau.edu/Tools/packrat_middens.htm.

Kious, W. Jacquelyne, and Robert I. Tilling. 2007. "This Dynamic Earth: The Story of Plate Tectonics." http://pubs.usgs.gov/gip/dynamic/dynamic.html.

Hoare, Robert. 1996–2008. WorldClimate. http://www.worldclimate.com. Climate data for many stations throughout the world.

5

Major Environmental Factors in Aquatic Biomes

Marine and freshwater environments are strongly controlled by factors related to the chemistry of water, although they are certainly not immune to the influences of regional climate or adjacent terrestrial environments. All waters on Earth are linked through the hydrologic cycle (see Figure 5.1). The oceans serve as major reservoirs of water, which is transformed from its liquid state to water vapor and released to the atmosphere by evaporation. Carried over land by winds and forced to rise and cool in a variety of ways (see Chapter 4), moisture in the atmosphere will condense into clouds. As tiny water droplets in the clouds collect, they merge into heavy raindrops, ice crystals, or snowflakes and fall to the Earth as precipitation. Upon the striking the land, the water either runs off as sheet flow or stream flow or percolates into the soil and becomes part of the soil moisture and groundwater supply. Gravity draws groundwater slowly toward sea level. It may return to surface flows as seeps and springs or enter streams, but whether above- or underground, it eventually returns to the sea. Diversions via evapotranspiration may shorten the loop that returns liquid water to atmospheric water vapor.

Properties of Water Influencing Lifeforms in Aquatic Biomes

Freshwater biomes offer different habitats and host different species than marine biomes, but many similarities exist in the physical conditions to which life must adapt. Some of these are described below and, where they exist, significant

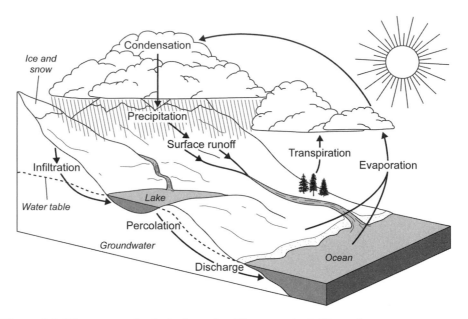

Figure 5.1 The water or hydrologic cycle. *(Illustration by Jeff Dixon.)*

differences between the two types of aquatic biomes are explained. The basic factors influencing life in water are interrelated, so discussion of one necessarily overlaps with discussions of others.

Transparency and the Penetration of Light

Compared with opaque land, water is transparent, allowing sunlight to penetrate surface layers. The longer wavelengths (red and orange) are absorbed within the top 50 ft (15 m) of the water column, a fact of more significance in the ocean than in much shallower freshwater streams, wetlands, and lakes. Most other wavelengths are absorbed in the next 130 ft (40 m), with the short-wave blue and violet light penetrating the deepest. These depths are greatly reduced when water clarity is reduced by particulate matter, either sediments or the microscopic organisms that make up the plankton. Many food chains in lakes and seas begin with single-celled cyanobacteria and algae that photosynthesize, so light is essential and its availability determines the depth to which primary production can occur as well as the timing of its seasonal pulses. Light also affects the ability of animals to sense prey and predators. Many zooplankters undergo daily and seasonal vertical migrations in their quest for food and invisibility.

In oceans, a threshold at which too little light prevents growth and reproduction of the phytoplankton occurs at a depth known as the compensation level. When only 1 percent of the sunlight reaching the surface remains, photosynthesis provides only enough energy for cell metabolism and nothing extra for growth or cell division. This depth marks the bottom of the euphotic zone, the uppermost layer in the vertical structure of deep lakes and oceans. The compensation level generally occurs about 650 ft (200 m) below the surface. Below this, only animals can live in the near-darkness. They depend on a rain of organic debris from above to sustain them.

Some faint light penetrates even deeper than the compensation level. Looking skyward, divers and marine animals can perceive a disk of light at a depth of 800 ft (250 m) and use it to navigate as they track the position of the sun overhead. Light overhead also provides a backdrop against which dark-colored fishes can be spotlighted, so countercoloration in which the dorsal side is dark and the ventral side is light or silvery is a common pattern in animals living in upper waters. Below 3,000 ft (1,000 m) light no longer exists. In the great oceans of the world, most habitat lies around 13,000 ft (4,000 m) below sea level, so darkness is the rule and light the exception.

When sunlight is absorbed by water molecules, it is converted to latent heat and sensible heat. The waters of the euphotic zone are therefore warmed and water temperature may fluctuate—sometimes markedly—with the seasons. Deep water is constantly cold.

Water as Universal Solvent

The water molecule, composed of two hydrogen atoms and one oxygen atom, is arranged in a nonlinear fashion that makes opposite ends of the molecule bear opposite charges; that is, water is a polar molecule (see Figure 5.2). The negatively charged electrons of the two hydrogen atoms are drawn toward the much larger

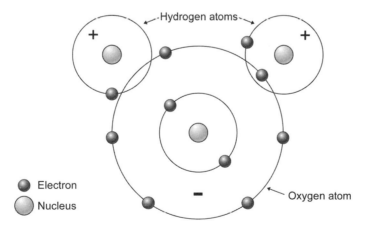

Figure 5.2 A water molecule, showing how the negatively charged electrons of the two hydrogen atoms are attracted to the outer ring of electrons of the oxygen atom, creating a polarized molecule. The hydrogen end bears a slight positive charge and the oxygen side a slight negative charge. This is one factor that makes water such a unique substance. *(Illustration by Jeff Dixon.)*

oxygen atom, leaving the hydrogen end of the molecule weakly positive and the oxygen end slightly negative.

Two consequences of polarity are (1) water molecules bind to each other, the reason it takes more energy to raise the temperature of water than other molecules, especially those forming land; and (2) water is able to rather easily dissociate molecules of salts or dissolve them. A salt is a solid that in water breaks apart into negative and positive ions, each attracted to one end of the water molecule. While common salt (sodium chloride or NaCl) is vastly abundant in the seas, other ions provide more important nutrients for phytoplankters. Calcium (Ca^{+2}), the sulfur in sulfates (SO_4^{-2}), nitrogen in nitrates (NO_3^{-1}), and phosphorus in phosphates (PO_4^{-3}) are essential to life.

Salinity is the measure of dissolved salts in water and today is often given in terms of practical salinity units (psu). The figure is actually a ratio and thus carries no unit of measurement. Average salinity of the sea is 35; fresh water has a salinity less than 0.5.

A highly saline environment (one greater than an organism's internal salinity) is challenging to most organisms, since water moves by osmosis from areas of low salt concentration to areas of high concentration. This imbalance threatens to draw water out of the cells and dehydrate them unless some physiological mechanisms exist to prevent water loss. The low salinity of freshwater can also be a problem, since water will want to pass from the environment into "saltier" living tissues.

The gases essential for life—oxygen (O_2), carbon dioxide (CO_2), and nitrogen (N_2)—also dissolve in water. These gases derive from the atmosphere and from living organisms. The capacity of water to hold dissolved gases is related to water temperature: the colder the water the higher the concentration of dissolved gas possible.

Much of the oxygen diffuses from the air, so surface layers of bodies of water tend to have the highest content. Microscopic phytoplankters and macrophytes release oxygen into the water as a by-product of photosynthesis, further enriching surface layers. Plants, animals, and microorganisms all use oxygen to respire, that is, to release the energy locked in organic compounds for their own metabolic use. Atmospheric carbon dioxide dissolves in water, which is one way that excessive loads of carbon dioxide pumped into the atmosphere by human activities may be somewhat reduced. Carbon dioxide is used during photosynthesis to manufacture organic compounds, but it is produced as a by-product of respiration. Thus, levels of oxygen and carbon dioxide fluctuate in water according to the daily and seasonal rhythms of biological activity.

Oxygen can be a limiting factor. Oxygen does not diffuse readily through water. Aquatic organisms absorb oxygen directly from the water and in the process can create a thin oxygen-depleted layer of water around their cells or bodies. To overcome this problem, they must constantly replace this film of water, either by evolving mechanisms or strategies to keep the water moving past them or by moving themselves. Elaborate gill structures in fishes and macroinvertebrates increase the surface area exposed to oxygen-laden water, and these animals usually flap or wave body parts to keep water moving past them.

Deeper water, a place devoid of photosynthesizers, can easily become depleted of oxygen, especially during times of high biological activity. Since the dissolved gas moves slowly on its own from surface to deep, some manner of mechanical mixing of the water layers is needed to replenish the supply. On larger lakes and in the ocean, wave action accomplishes this. Deep vertical currents in the seas transport cold, oxygen-rich polar water to the sea floor, so the bottom waters are usually well-oxygenated. (Water at intermediate depths, however, may have low concentrations of oxygen.) This is not the case in lakes, which lack currents. Very deep lakes, such as those in Africa's Rift Valley, are lifeless in their deepest waters due to lack of oxygen. Shallower lakes in the mid-latitudes replenish deep water oxygen levels during the seasonal turnovers that result from changes in water temperature and density, which is discussed next.

Water Density and Temperature

Another important aspect of water chemistry for life in aquatic biomes is water density, the mass of water per unit volume, which at sea level is about 800 times greater than the density of air. As a consequence of its density, water provides physical support for organisms, and aquatic animals do not need the skeletal strength of terrestrial animals simply to support their bodies. Large plants growing in shallow waters do not need strong woody trunks and branches. On the other hand, animals have to be able to move through water, something more difficult to do than moving through air. Streamlined bodies and strong musculature attached to fins or tails help propel them through this relatively viscous medium. Still, most organisms are denser than water and so tend to sink. Various structures and behaviors have evolved to increase buoyancy and help them hold their position in the water column or rise and fall as needed. Some phytoplankters are aided by their shape, as any departures from a spherical shape works to retard sinking. The amount of gas in vacuoles in the cell can make an alga less or more dense. Many protists have flagella or other appendages that help propel them up and down the water column. Many fish have swim bladders they can fill and empty with internally generated gases.

The density of water changes with temperature. Warm water is less dense than cold water and will float on top of it. This is a major reason that the water column of lakes and oceans becomes stratified during warm seasons in the mid-latitudes and year-round in the tropics. Beneath the warm surface layer, temperature decreases rapidly in a fairly narrow transition zone called the thermocline. Below the thermocline, water temperature varies little with increasing depth. In the ocean, deep sea temperatures remain at 37° F (3° C) all year, except in the deepest areas where they may be between 33° and 35.5° F (0.5° and 2.0° C). Just before freezing (32° F or 0° C in freshwater; 28.5° F or −1.9° C in seawater), water density suddenly become less and the coldest water floats up to the surface. Ice formation begins at the top of the water column in lakes and seas, not at the bottom.

Since temperature changes rapidly with depth just below the warmed euphotic zone, so does density. A pycnocline or zone of rapid density change separates the surface waters from those denser waters of the deep sea and can prevent nutrients that sink below it from rising back into the euphotic zone where they could be used by the phytoplankters. Strong mixing at fronts, by storm waves, or by upwelling overcome the barrier of the pycnocline and create some of the most productive regions in marine biomes.

In the ocean, salinity also influences density. The more saline the water the greater its density. High evaporation rates in the tropics and during mid-latitude summers increase the density of surface waters. High precipitation totals, abundant stream runoff, and snow- and icemelt lower the density of surface waters. Less saline water floats on the denser water of greater salt content. The transition zone between the two layers is called a halocline. The conveyor belt of deep ocean circulation seems to be driven by salinity and density gradients. Warm Gulf Stream water from the tropics and subtropics is saltier and hence denser than the surface water of North Atlantic Ocean, so it sinks, pulling more water down behind it. At the onset of the Little Ice Age, according to one hypothesis, the conveyor belt stopped because fresh meltwater from the retreating ice sheet over North America spilled down the St. Lawrence River valley and into the Atlantic Ocean, lowering the salinity and hence density of the surface waters of the ocean.

The stable layers that comprise a stratified water column dissipate in autumn and winter as the surface waters lose their warmth and as bottom waters near freezing rise to the surface. This process of turnover mixes the layers in many mid-latitude lakes and seas and brings nutrients back to the euphotic layer.

Water in Motion

One of the key differences between land and water is that water moves. This is important in the transfer of heat energy; in nutrient recycling; in the transport of seeds, plants, eggs, and larvae; and in erosional and depositional processes shaping shorelines and benthic habitats.

The lotic environment of rivers and streams is the most dynamic of freshwater biomes. Flowing water shapes the channel, sometimes eroding the banks and stream bottom and sometimes depositing sediments. The volume of water in a stream changes seasonally and after major precipitation events. As the volume changes so too does the velocity of the flow. The dominant movement is, of course, downstream. Nutrients, sediments, and organisms are all swept along. Aquatic life-forms develop ways to resist this force or work with it. They may cling to the rocks or bury themselves in channel gravels and sands. Flattened bodies let water pass right over or by them. Insect larvae may wash downstream, but adults can fly back upstream.

In lakes and oceans, waves stir up the bottom and wash or pound the shore. The turbulence created can resuspend particles, including phytoplankters and their nutrients, in the euphotic zone. Waves flooding up a beach dislodge grains of sand

and move them along the shore. The energy of a surf erodes the shore, and particles held in suspension abrade rock and the shell and exoskeletons of invertebrates.

In oceans and coastal wetlands, both freshwater and saltwater, the motion generated by tides is significant in the daily and seasonal patterns of life. Alternately exposing and inundating the seabed, tides demand special adaptations of species living in the intertidal zone for them to withstand both extremes. Egg laying and hatching by a number of species may be timed to coincide with spring tides, the highest and lowest tides of the month. Tidal currents redistribute sediments in estuaries and help produce the mudflats that support numerous invertebrates and vertebrates.

Most significant in the global environment is the movement of the great ocean currents. Surface currents transport heat from the equatorial zone poleward and distribute it to higher latitudes, preventing excessive heat buildup in the tropics and moderating temperatures in the middle and high latitudes. They influence not only maritime climates but continental ones, as well. The ocean currents are driven by the global wind system and deflected into huge gyres by continents. The centers of the gyres lie beneath subtropical high-pressure cells in the global atmospheric

....

The Great Pacific Garbage Patch

The debris of civilization slowly rotates around the center of the North Pacific subtropical gyre. The permanent subtropical high-pressure cell located at 30° N depresses the ocean surface and so draws in particles drifting in the encircling ocean currents. Although the high pressure produces the strong winds that cross the Pacific basin, at the center of the cell, winds are notoriously calm, much too light to mix the water column, so garbage stays afloat for years with nowhere to go but round and round an area twice the size of Texas. Some of the junk is recognizable as plastic shopping bags and bottles, fishing lines, and even an occasional traffic cone, but much consists of plastic pellets or particles of Styrofoam®.

Surface waters at the center of gyres are naturally low in phytoplankton and therefore other forms of life. Filter-feeding zooplankters such as salps readily ingest the plastic; colored pellets can be seen easily in their transparent bodies. At the other end of the food chain, albatrosses "catch" plastic flotsam and other debris to feed their nestlings. Like sea turtles that clog their intestines with pellets, six-pack rings, and baggies, the young birds starve because this kind of garbage has no nutritive value.

Other threats to sea life posed by the floating garbage include entanglement and poisoning. Plastic pellets tend to attract and concentrate toxins such as DDT and PCBs. When salps and jellyfishes are eaten by fish, the toxin is magnified and passed in ever higher concentrations up the food chain—possibly even to human consumers.

All gyres, some 40 percent of the ocean surface, collect our throwaways. Plastics will degrade eventually in sunlight or through oxidation and be recycled as inorganic nutrients, but no one knows how long this will take. Some scientists estimate 500 years or more.

....

circulation system, the areas out of which winds spiral. Flotsam travels the world on the currents and plants and animals float or hitch rides along the way. The strong easterly Trade Winds of the tropics push warm ocean waters to the west until they are diverted north or south along a continental margin. Hence, the western boundary waters of oceans are typically warm, especially in the tropics and subtropics. The strong steady flow of the Trade Winds pushes surface water away from the western coasts of continents. Cold water from below the thermocline takes its place and produces cold currents along the eastern sides of ocean basins. The cold upwelling waters bring nutrients to the surface and support great fisheries upon which seabirds and marine mammals often feed, not to mention human populations.

Deep oceanic circulation occurs as a slow surface-to-sea-floor flow. Cold dense water sinks off the coast of Antarctica and in the Arctic and North Atlantic oceans and then moves along the bottom of the oceans in a single huge global traverse. The waters rise back to the surface at major areas of upwelling. It is estimated that it takes a molecule of water about 2,000 years to make the full circuit.

Zonation in Aquatic Environments

A common characteristic of both freshwater and marine aquatic ecosystems is the development of distinct habitat belts or life zones inhabited by unique sets of organisms (see Figure 5.3). Vertical zones reflect water depth and the difference between

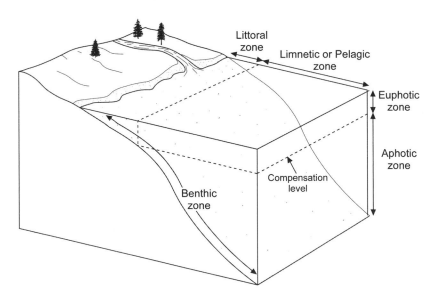

Figure 5.3 Lakes and oceans are both characterized by zonation of environmental conditions and hence life. The details vary, but both have a similar arrangement of zones. *(Illustration by Jeff Dixon.)*

pelagic (open water) and benthic (channel, lake bottom, or seabed) communities; horizontal zones reflect differences encountered with distance from shore. Variations are introduced by differing substrates, particularly soft sediments versus hard rock surfaces.

Freshwater Biomes

Longitudinal zonation is a feature of river systems, a pattern recognized in the River Continuum Concept, which summarizes the changes in physical characteristics occurring in the downstream progression from small headwater streams of the uplands to mid-size streams to the large rivers of the lowlands. Slope, volume of flow, water depth, sizes of particles that make up the stream's load, turbidity, channel breadth, inputs of terrestrial materials, water temperature, and the amount of shade versus direct sunlight all affect the amount of instream primary productivity and thus the biota and food chains in the different parts of the stream system. A vertical component to zonation is developed in the hyporheic zone, the water-saturated substrates next to the channel bed and sides where groundwater and river water come together.

In lakes, as in oceans, the vertical and horizontal zones merge to some extent. The most fundamental division is between the benthic zone (the lake bottom sediments) and water habitat itself. The latter changes with distance from shore and the depth of the water. The near-shore zone or littoral zone consists of shallow waters where light and oxygen are available. It can be variously defined as the area between low water and high water (and therefore similar in concept to the littoral zone in the Marine Biome), the zone where rooted emergent plants may grow, or the zone where light is adequate for the growth of submerged aquatic vegetation. Substrate, whether uncondolidated sediments or solid rock, influences the actual types of organisms occupying a lake's littoral zone.

Beyond the influence of the shore lies open water or the pelagic zone. Vertical layers develop in this zone as a consequence of temperature differences. The warmed surface waters make up the epilimnion. Below that layer is the thermocline, a layer in which temperature decreases rapidly with depth. The uniformly cool waters below the thermocline are referred to as the hypolimnion. In deep lakes, another zone occurs beyond the influence of sunlight. This is known as the abyssal or profundal zone.

Marine Biomes

As with lakes, the habitats of the sea can be coarsely subdivided into a benthic zone and a water environment, and the water habitats vary with distance from shore and depth below the sea's surface. The major components of the physical environment that lead to distinct zones in the ocean include temperature, pressure, currents, waves, tides, and nutrient input. For the benthic zone, the main physical controls relate to exposure to wave action and particle sizes in the substrate.

The coastal or littoral zone is defined as the area between the high-tide mark and the low-tide mark. Also known as the intertidal zone, all parts are alternately exposed to the atmosphere and inundated by saltwater for varying lengths of time, usually on a twice-daily basis. During exposure, marine life faces direct sunlight, high and rapidly changing temperatures, and the threat of drying out. During periods of inundation, the environment becomes waterlogged and wave action and longshore currents may become powerful. Holding fast can become a problem. Since tidal heights vary through the month and sometimes each day, vertical subzones are recognized in the coastal zone because the physical characteristics of each area results in the dominance of a different set of organisms. The supralittoral zone exists on rocky shores above the true coastal zone. Seaspray from waves crashing on the rocks below creates a zone in which only salt-tolerant land plants and associated animals can survive. The eulittoral zone is the true intertidal area. A subtidal zone exists in shallow water never truly exposed but affected by different water depths as the tide ebbs and flows.

Beyond the low-tide mark is the pelagic zone or open sea. Distinction is made horizontally between the shallow waters overlying the continental shelf, the neritic zone, and the waters of the deep sea. Vertical zonation becomes important as water deepens in the pelagic zone. The epipelagic zone corresponds to the euphotic zone, the zone through which sunlight adequate for photosynthesis penetrates. The zone extends to depths of approximately 650 ft (200 m) below the sea's surface. Beyond the continental shelf, four deeper zones occur. The mesopelagic extends from the epipelagic to a depth of about $-3,200$ ft $(-1,000$ m). The bathypelagic zone lies between $-3,000$ and $8,000$ ft $(-1,000$ to $-2,500$ m). The abyssopelagic zone makes up almost all the remaining deep water and reaches from $-8,000$ ft $(-2,500$ m) to within 300 ft (100 m) of the sea floor. To accommodate the extreme conditions found in oceanic trenches, a subzone of the abyssopelagic, the hadal zone, is recognized for waters deeper than $-20,000$ ft $(-6,000$ m). The final zone in ocean waters is the benthopelagic zone, which contains the layer of water lying within 300 ft (100 m) of the sea floor. This is a separate life zone because larvae and other forms of benthic life frequently rise into these bottom waters.

Latitudinal Influences

Latitude determines the length and extremes of seasons. Its strong influence on annual temperature patterns means it also determines freeze-thaw cycles of pack ice and the timing of the mixing of layers (or general lack thereof) in lakes and ocean waters. General distinctions can be made in all aquatic biomes between polar, warm temperate, cool temperate, and tropical conditions. In the sea, the strict relationship between latitude and temperature is modified along coasts by ocean currents. Warm currents can extend tropical and subtropical temperatures poleward; cool currents lower temperatures in these same latitudes.

The connection between latitude and global atmospheric circulation patterns (see Chapter 4) means that latitude also has an influence on precipitation patterns and therefore salinity and water density in lakes and seas and flow regimes in rivers and streams. Water levels in lakes and wetlands vary with precipitation and runoff. The formation of ice is a significant disturbance factor on polar coasts and creates vast wetlands on Northern Hemisphere lands where northward flowing rivers remain blocked by ice in their lower reaches at a time when tributaries upstream have already thawed.

Change Through Geologic Time

Plate Tectonics

Plate tectonics have greatly influenced the marine biomes. Sea-floor spreading and related shifts in the positions of continents on the globe have created and destroyed habitats and altered currents and water temperatures. Mid-oceanic ridges mark the location of active sea-floor spreading at the margins of plates and have hydrothermal vents associated with them. Hot spots occur away from plate boundaries where for reasons largely unknown plumes of magma rise from the mantle. Plates move slowly over these hot spots, lingering long enough for huge shield volcanoes to form a string of volcanic islands. As volcanoes on the plate move beyond the active plume, they become extinct and lose the buoyancy provided by the magma chamber once beneath them. Tropical islands that had been fringed with coral reefs disappear beneath the sea and coral atolls mark their former locations. After millions of years the extinct volcanoes sink far below the sea's surface and become seamounts. If the summits remain near the surface long enough, wave action may flatten them and form guyots. The Hawaiian Islands and the long chain of seamounts extending northwest of them were formed in such a way. Where one oceanic plate is subducted under another oceanic plate, such as east of Guam, or under a continental plate, such as along the west coast of South America, oceanic trenches form (see Figure 5.4).

The convergence of continental masses to form a single supercontinent by the end of the Paleozoic Era would have eliminated a lot of shallow water habitat as coastlines and continental shelves became part of the land. At the beginning of the Mesozoic Era when the supercontinent Pangea began to split apart, ocean basins began to change and slowly develop their present configuration (see Figure 5.5). A basin that had been nearly surrounded by Pangean land opened and spread westward between Gondwana and Laurasia (see Chapter 4) to form a sea that essentially encircled the globe. This Mesozoic seaway is called the Tethys Sea. Mangroves lined its tropical shores.

Later, in the Miocene Period, the Tethys Sea would close as the African-Arabian plate became sutured to Eurasia and leave only the Mediterranean Sea as a sign of its past existence. The exchange of water and organisms between the Pacific and Indian oceans was cut off some 19–20 million years ago by the northward progression of the Australian–New Guinea Plate. From the late Miocene into the Pliocene, some 3–5 million years ago, modern atmospheric and oceanic circulation

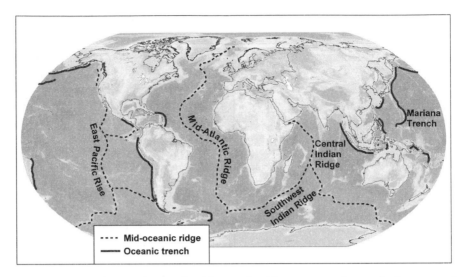

Figure 5.4 Oceanic trenches and mid-oceanic ridges are products of plate tectonics (compare with Figure 4.12). They offer different habitats for marine life. *(Map by Bernd Kuennecke.)*

was established. Several changes in the arrangements of oceans and landmasses were responsible. The Bering Strait opened and connected the Arctic and Pacific Oceans; circulation in the Indian Ocean was essentially confined to that ocean basin; and the Drake Passage between South America and Antarctica opened,

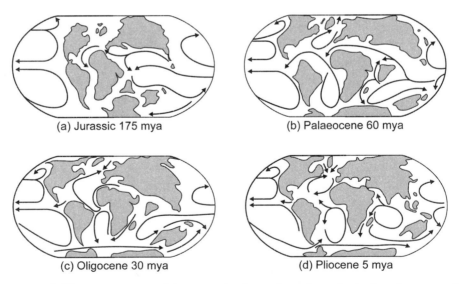

(a) Jurassic 175 mya
(b) Palaeocene 60 mya
(c) Oligocene 30 mya
(d) Pliocene 5 mya

Figure 5.5 The movements of the tectonic plates through geologic time have greatly altered ocean currents, which in turn have greatly affected climates and the distributions of organisms on land and in the sea. *(Illustration by Jeff Dixon.)*

allowing a circumpolar cold ocean current to dominate the Southern Ocean and effectively separating tropical and polar waters in the Southern Hemisphere. The final step was the completion of the Central American isthmus 3 million years ago, which divided the waters of the Atlantic from those of the Pacific. On land, the major mountain ranges that influence atmospheric patterns were in place by then and west coasts in the subtropics had become arid. The new relationships between land and sea provided permanence to subtropical high-pressure cells over the oceans; and they, in turn, through the winds blowing from them, provided the energy for the great gyres of surface currents that dominate on either side of the Equator. Wind patterns created upwelling zones and cool surface waters on the eastern borders of ocean basins and the warm currents that move heat poleward along the western borders.

Incomplete separation of plates occurred in several locations during the break up of Pangea. Deep rift valleys formed as blocks of continental crust dropped into space created by diverging plates, but the process halted before new sea floor could develop. The Great Rift Valley of East Africa is the best and largest example of this type of landform, and in several places its basins contain ancient and deep lakes, the largest of which are Lake Tanganyika and Lake Malawi. The world's deepest lake and largest body of freshwater, Lake Baikal, also lies in a rift valley. Other lakes occupying depressions formed as a consequence of plate tectonics include Lake Titicaca, high in the Andes; Lake Tahoe, lying in a trough between the uplifted Sierra Nevada to its west and the Carson Range to its east; and Lake Vostok, buried under the ice cap of Antarctica for nearly 15 million years.

Pleistocene Changes

The climate swings of the Pleistocene Epoch that alternated repeatedly between cold glacial periods and warm interglacial periods had lasting effects on marine biomes, especially those of the coast and continental shelf. Sea surface temperatures and sea levels changed. Sea level had dropped on average 400 ft (135 m) during glacial maxima, exposing large areas of the shelf as dry land. Rivers could now carve valleys across new coastal plains. It is estimated that today about 16 percent of the continental surface rocks are submerged as continental shelf. During the height of glaciation in the late Pleistocene, only about 5 percent remained below sea level. Major changes in marine communities, with species displaced or going extinct, had to have occurred. While geologically most shelf areas are approximately 200 million years old, their biological communities are only 10,000 years old or younger.

The warming of the Holocene raised sea level as meltwaters drained to the oceans. As the sea rose again, river valleys on Pleistocene coastal plains were flooded by the sea and became estuaries. Deep troughs carved by mountain glaciers running to the sea in the high latitudes of western Canada, Norway, and Chile were also inundated and became fjords. Thick glacial deposits laid down between the United Kingdom and Scandinavia were submerged and became banks, as happened to similar deposits off the east coast of North America. Most coral reefs were drowned, so modern communities were likely assembled less than 10,000 years ago.

Freshwater aquatic biomes were also affected by the climate changes of the Pleistocene. Ice blocked some rivers and caused ponding or course changes. For example, mountain glaciers dammed the Truckee River, Lake Tahoe's only outlet; as a consequence, lake waters rose to a level 800 ft (240 m) higher than today. Wetter climates prevailed in many of today's semiarid and arid regions, and rivers flowed and vast lakes formed in interior basins. Lake Manly flooded Death Valley, Lake Bonneville covered much of western Utah, and hundreds of glacial lakes dotted the Great Basin. With the retreat of great ice sheets, ice dams failed and the Pleistocene lakes drained in torrents that eroded the land into deep wide valleys through which today only small rivers may flow. As climates south of the ice sheet became drier, the glacial lakes evaporated, leaving salt flats, playas, and salt lakes in their place. Free of ice blockage, some rivers assumed new courses. The Teays River that had flowed from North Carolina and Virginia into the Mississippi now found an easier outlet to the ocean through the St. Lawrence Valley and cut away from its main course. The part of the channel that had been directed west by the edge of the ice became the Ohio River.

As the ice retreated in the Holocene, thick deposits of glacial drift impeded drainage and led to the formation of myriad ponds and bogs. Most of the world's lakes exist in areas formerly covered by ice and are less than 10,000 years old. In other areas, the ice had scoured the bedrock and gouged out great depressions. Exposed again to the atmosphere, these depressions filled with water and became features such as the Great Lakes of North America and the Finger Lakes of New York.

Short-Term Changes

Most freshwater and some marine habitats are in near-constant states of flux. Rivers alternately erode or deposit sediments in their channels depending on the volume and velocity of water flowing in them. Undercut banks collapse; sand bars shift position. The bed load tumbles along, so the channel bottom is often an unstable substrate. During periods of high water, rivers overflow their banks and spread sediments across the floodplains they are building. On great rivers like the Amazon and Yangtze, floodplain lakes reform each year as the floodwaters recede. Rivers frequently change course and erode new channels across the floodplain when above flood stage. Meandering rivers cut off segments of the channel to form short-lived oxbow lakes.

All lakes eventually fill in as sediments wash in from the land surrounding them and dying marsh and swamp plants contribute organic matter. The process may happen in a hundred years or less in some oxbow lakes, but it can take millions of years in a deep rift valley lake. Wetlands too are filled and then created anew elsewhere. The acidic nature of bogs holds them in a state of suspended animation for a longer time than most other types of freshwater wetlands.

The sands and muds on soft sediment coasts are moved constantly with the swash and backwash of the waves. Burrowing animals, deposit feeders, and those that probe the soft sediments for food constantly churn the bottom materials. Salt marshes and mangroves trap sediments eroded from the land and reshape the substrate, building it up above the reach of saltwater and drying it out enough that the habitat no longer supports them as well as it does truly terrestrial vegetation.

On a somewhat different time scale, deep sea hydrothermal vents can be considered short-lived features. As the sea floor spreads away from the mid-oceanic ridges where they were heated by rising magma, the rising plume of superheated water rich in dissolved minerals stops. New vents appear on younger oceanic crest. How vent animals get from one vent to another is still a mystery, but one hypothesis is they can use randomly placed carcasses of whales and other large forms of marine life as stepping stones. This dead organic material may decay slowly in the cold, dark depths of the ocean, but nonetheless any carcass is an ephemeral feature.

Aquatic Biomes and Contemporary Global Climate Change

The lessons of the past inform the present and contribute to the formulation of models that will predict the future. Study of the Pleistocene, in particular, has provided valuable insights into the responses of the physical environment as well as biological communities to climate change. The interconnections of atmospheric and oceanic systems, of climate and the hydrologic cycle, of greenhouse gases and air and sea surface temperatures all point to significant changes in aquatic biomes as the global climate changes. Many unknowns remain; the oceans are barely explored and the complexity and interconnectedness of atmospheric processes still begs adequate understanding. Global climate change models are necessarily imprecise. Nevertheless they point in directions that cannot be ignored, and some impacts of global warming are already apparent.

On average, global temperatures have warmed 1° F (0.6° C) in the last 100 years. Much of the increase occurred in the last 30 years. The warming is not uniform. The fastest rise in temperatures has occurred in the Arctic, where a shrinking ice cover accelerates warming by reducing the amount of sunlight reflected back to space. Temperatures in the Arctic Ocean were 1.8°–7° F (1°–4° C) higher in 2006 than in 1976. Much greater warming has occurred in the Northern Hemisphere than the Southern Hemisphere because of the preponderance of land in the former and the moderating effect of the vast seas in the latter. Temperatures in the Indo-West-Pacific region of species-rich coral reefs has risen only 0.4°–1.8° F (0.2°–1° C), still an alarming rise in such a sensitive part of the world ocean. Even a slight rise in water temperature is fatal to many coral reef organisms. Widespread coral bleaching, in which stressed coral polyps expel the symbiotic xooanthellae living in their tissues and giving them color, is already reported. Sometimes the

•••

Sulfur Events in the Waters off Namibia

The Namib Desert is one of the driest places on the Earth, but rare rain events may now be accompanied by a whitening of coastal waters and the distinct odor of hydrogen sulfide gas. One of the world's major upwelling zones lies offshore. Nutrients brought to the surface result in a highly productive phytoplankton that once supported billions of sardines. The fishery was severely depleted in the 1970s by commercial fishing fleets from other parts of the globe and never recovered. It appears that the loss of the major herbivore in the marine ecosystem changed the system, and most phytoplankters go unconsumed, die, and settle to the bottom, where they decay. Anaerobic decay processes produce hydrogen sulfide gas and methane gas that normally are held in the seabed sediments by the high atmospheric pressure that dominates the region. When a small, traveling low-pressure system moves westward across southern Africa, it not only brings rain, but also acts like an open pressure valve and releases the gases, which rise to the surface.

Hydrogen sulfide gas stinks and the odor of rotten eggs wafts over the coast. It is also toxic and thousands of dead fish wash up on the beaches near Swakopmund, adding their stench to the brew. The gas reacts with water to form pure sulfur, giving the sea a milky cast.

Some scientists think a positive feedback loop has been set up by the methane that escapes to the atmosphere during these so-called sulfur events. Methane is a much more powerful greenhouse gas than carbon dioxide and a major contributor to global warming. With increased warming may come stronger winds driving stronger upwelling and the ever-greater production of plankton and its by-products—hydrogen sulfide, smelly fish kills, more methane, and more warming.

•••

xooanthellae and corals get together again, but coral bleaching can spell the end of shallow water reefs.

Atolls and other low-lying islands and coasts are already experiencing the effects of rising sea levels. Even without significant melting of the Greenland and Antarctic icecaps, warming water expands and this accounts for the rise witnessed so far. Melting sea ice does not directly affect sea level, but is evidence of the warming now occurring in polar latitudes and can create a host of other problems, including the loss of feeding or breeding grounds for marine animals, such as polar bears, whales, and walruses in the Northern Hemisphere and certain kinds of penguins in the Southern Hemisphere, not to mention possible territorial disputes between nations with economic or defense interests in areas increasingly free of the barrier of ice.

Increased storminess is also probable on a warmed Earth. Together with higher sea levels, more frequent and stronger storms threaten coasts with accelerated erosion and flooding by seawater. Coastal wetlands, some of the planet's most productive ecosystems, can be destroyed.

A warmer atmosphere will be able to hold more water vapor. This fact leads to predictions of increased evaporation rates and aridity in some regions of the Earth and increased cloudiness and precipitation in others. All models do not yet agree on just where these effects will take place. It can be expected that some lakes will shrink and others expand, some rivers may dry up and others will flood more often. Since all of this is happening at a time of rapid growth of the human population and increased demand for potable water as well as water for industrial use, great hardship and major population relocations may be part of the human experience in the near future.

Increased cloudiness could actually reduce global temperatures by reflecting incoming sunlight back to space. Decreased ice, one the other hand, will reduce the albedo or reflectivity of polar area, and the greater absorption of solar radiation will contribute to warming the lower atmosphere. These are just two of the contradictory responses of air, water, and ice that make predictions so difficult and the future risky. The problem of global change is not to be ignored but should be seen as an urgent call for research and planning for dealing with possible impacts.

The oceans are major sinks of carbon dioxide. They take carbon dioxide—a major greenhouse gas—from the atmosphere and store it long term in the calcium carbonate shells of the zooplankton, which upon death settle to the sea floor. It has been estimated that about half of the CO_2 added to the atmosphere by the burning of fossil fuels since the early industrial era has been absorbed by the ocean. The current rate of absorption is some 22 million tons a day. There appear to be limits to the ability of the seas to absorb excess CO_2 from the air, however, and some consequences. Some of the dissolved CO_2 forms carbonic acid (H_2CO_3) in water. Hydrogen ions (H^+) from the acid bind more readily to carbonate ions (CO_3^{-2}) than to calcium ions (Ca^{+2}) and thus an increase in hydrogen ions interferes with the production of calcium carbonate by organisms that use it to build the structures that encase them. A measurable drop in the pH of the sea has occurred already, and forecasts are for even-greater reductions by 2100. The process is called ocean acidification, although the oceans will not actually become acidic. The neutral point on the pH scale is 7.0. The ocean's pH has already declined from 8.2 to its current level of 8.1; in other words, it is shifting toward the neutral point or toward the acid end of the scale. By the end of this century, pH could drop to 7.8 or even 7.7. Carbonates available to calcifying organisms have already declined 16 percent since the beginning of the industrial age in some warm water regions.

Acidification of the oceans affects life at all levels. Many corals could become extinct, and with their loss shallow water reefs, the habitat for so many marine species, will disintegrate. Reef-building coralline algae are similarly threatened. In the geologic past, during the Triassic Period, increases in CO_2 coincided with the loss of many cnidarians; survivors were the non-reef-building anemones. Today, CO_2 concentrations in the atmosphere are increasing at rate 1,000 times greater per century than at any time in the last 420,000 years. The fear is that this may be much too fast for evolution to keep pace, so extinctions may be more widespread that during the Triassic. Reef-building corals are just one concern. With lower pH, abnormal growth occurs in larval sea urchins, oysters, pteropods (zooplankters), and coccolithophores, which are members of the phytoplankton. Coccolithophores

Figure 5.6 Coccolithophore. The calcium carbonate plates that coat the surfaces of these algae will be more difficult to produce if the oceans continue to acidify. *(Illustration by Jeff Dixon.)*

are one-celled algae coated by more than 30 plates composed of calcium carbonate (see Figure 5.6). Their spring blooms can be so extensive that they turn surface waters a milky turquoise visible from space.

Coccolithophores generally inhabit nutrient-poor subpolar waters and are the ocean's predominant calcifiers. Their demise could change grazing food chains in large areas of the ocean. Not all organisms are negatively affected by acidification. An environment richer in dissolved CO_2 favors cyanobacteria, nitrogen-fixers that are common in the open sea.

Not all parts of all oceans are warming. Some regions, especially where cold deep water is upwelling, are actually getter cooler by as much as 2° F (1°–2° C) or even more. Most regions of cooling are in the Southern Hemisphere. In the South Pacific, cooler temperatures are being recorded all along the west coast of South America and in equatorial waters stretching past the Galapagos Islands and thousands of miles to the west. Very cold water appears at several locations in the Southern Ocean off the coast of Antarctica. Yet the waters surrounding the Antarctic Peninsula are among those warming the fastest, and as a consequence, enormous chunks of the ice shelves attached to its coasts continue to break off.

Further Readings

Books

Roth, Richard A. 2008. *Freshwater Aquatic Biomes.* Greenwood Guides to the Biomes of the World. Westport, CT: Greenwood Press.

Woodward, Susan L. 2008. *Marine Biomes.* Greenwood Guides to the Biomes of the World. Westport, CT: Greenwood Press.

Appendix

Representative Climate Data for Terrestrial Biomes

The following tables show average monthly and total annual precipitation, in inches, and annual monthly temperatures in degrees Fahrenheit. Climographs for some of these stations can be found in Chapter 4 and in the volumes of Greenwood Guides to the Biomes of the World that deal with the respective biomes.

Tropical Rainforest Biome

Andagoya, Colombia   Tropical wet (Af)
05° 06′ N, 76° 40′ W; elev. 250 ft

	J	F	M	A	M	J	J	A	S	O	N	D	Yr.
Precip. (in)	25.4	21.8	19.8	26.5	25.9	26.2	23.9	23.9	25.0	23.0	22.8	19.8	260.1
Temp. (°F)	82.0	81.8	82.8	82.7	81.6	80.6	81.4	81.4	81.5	81.7	81.0	81.0	

Belem, Brazil   Tropical wet (Af)
01° 23′ S, 48° 29′ W; elev. 53 ft

	J	F	M	A	M	J	J	A	S	O	N	D	Yr.
Precip. (in)	13.8	16.2	17.4	14.6	11.1	6.5	6.1	4.8	5.1	4.1	4.0	7.9	110
Temp. (°F)	81.0	80.0	80.0	81.0	82.0	82.0	82.0	82.0	82.0	82.0	82.0	82.0	

Douala, Cameroon   Tropical wet (Af)
04° 00′ N, 09° 44′ E; elev. 30 ft

	J	F	M	A	M	J	J	A	S	O	N	D	Yr.
Precip. (in)	2.0	3.2	7.7	8.9	12.1	18.8	24.5	24.7	22.9	16.5	6.1	2.2	150.1
Temp. (°F)	81.0	82.0	82.0	81.0	80.0	79.0	77.0	77.0	77.0	78.0	80.0	81.0	

Madang, Papua New Guinea   Tropical wet (Af)
05° 13′ S, 145° 47′ E; elev. 14 ft

	J	F	M	A	M	J	J	A	S	O	N	D	Yr.
Precip. (in)	13.3	11.5	13.6	17.4	14.8	8.0	6.7	5.0	5.7	11.3	16.1	15.5	138.8
Temp. (°F)	80.0	80.0	80.0	80.0	80.0	80.0	79.0	80.0	80.0	80.0	80.0	80.0	

Tropical Seasonal Forest Biome

Minantitlan, Mexico   Tropical wet and dry (Aw)
17° 59′ N, 94° 31′ W; elev. 90 ft

	J	F	M	A	M	J	J	A	S	O	N	D	Yr.
Precip. (in)	3.4	2.5	1.5	1.9	3.7	9.4	10.7	12.1	20.0	14.9	17.6	4.1	101.9
Temp. (°F)	71.0	73.0	76.0	78.0	81.0	80.0	80.0	79.0	79.0	77.0	73.0	71.0	

Kananga, Democratic Republic of the Congo   Tropical wet and dry (Aw)
05° 53′ S, 22° 28′ E; elev. 2,146 ft

	J	F	M	A	M	J	J	A	S	O	N	D	Yr.
Precip. (in)	4.7	4.5	7.3	6.1	3.2	0.5	0.7	2.0	4.6	5.8	9.2	8.1	56.7
Temp. (°F)	76.0	76.0	77.0	77.0	77.0	76.0	74.0	76.0	76.0	76.0	76.0	76.0	

Akyab, Myanmar   Tropical monsoon (Am)
20° 01′ N, 92° 54′ E; elev. 16 ft

	J	F	M	A	M	J	J	A	S	O	N	D	Yr.
Precip. (in)	0.1	0.2	0.4	1.5	13.8	45.6	50.7	42.4	23.1	11.3	3.7	0.6	193.5
Temp. (°F)	68.0	71.0	77.0	82.0	82.0	80.0	78.0	78.0	80.0	80.0	77.0	71.0	

Petrolina, Brazil (caatinga)   Semiarid (BSh)
09° 38′ S, 40° 40′ W; elev. 1,233 ft

	J	F	M	A	M	J	J	A	S	O	N	D	Yr.
Precip. (in)	2.6	3.1	3.9	1.9	0.4	0.2	0.1	0.1	0.1	0.5	1.8	2.5	17.0
Temp. (°F)	80.8	80.6	79.7	78.8	77.9	76.1	75.0	76.5	79.7	81.0	82.6	82.0	

Tropical Savanna Biome

Kano, Nigeria   Tropical wet and dry (Aw)
12° 02′ N, 08° 31′ E; elev. 1,563 ft

	J	F	M	A	M	J	J	A	S	O	N	D	Yr.
Precip. (in)	0.0	0.0	0.1	0.4	2.5	4.6	7.9	11.3	4.9	0.5	0.0	0.0	32.2
Temp. (°F)	70.2	74.8	82.0	87.6	87.3	83.3	79.2	77.7	79.2	80.4	76.3	71.6	

Nairobi, Kenya   Tropical wet and dry (Aw)
01° 19′ S, 36° 55′ E; elev. 5,327 ft

	J	F	M	A	M	J	J	A	S	O	N	D	Yr.
Precip. (in)	2.2	2.0	3.0	7.8	6.3	1.6	0.6	0.7	1.0	2.0	5.1	3.3	35.6
Temp. (°F)	64.6	66.2	66.2	64.8	63.0	60.8	59.5	59.5	62.2	64.0	63.5	63.3	

Cuiaba, Brazil   Tropical wet and dry (Aw)
15° 38′ S, 56° 06′ W; elev. 623 ft

	J	F	M	A	M	J	J	A	S	O	N	D	Yr.
Precip. (in)	10.0	8.6	8.1	4.0	2.2	0.5	0.3	1.1	1.8	5.2	6.0	8.1	55.9
Temp. (°F)	81.0	82.0	81.0	80.0	77.0	75.0	75.0	78.0	81.0	82.0	81.0	81.0	

Desert Biome

Warm Desert

Hyderabad, Pakistan   Warm desert (BWh)
25° 22′ N, 68° 25′ E; elev. 135 ft

	J	F	M	A	M	J	J	A	S	O	N	D	Yr.
Precip. (in)	0.1	0.2	0.2	0.1	0.1	0.4	2.8	2.3	0.7	0.0	0.1	0.1	7.1
Temp. (°F)	62.0	68.0	78.0	86.0	91.0	93.0	89.0	87.0	86.0	84.0	75.0	66.0	

Gila Bend, Arizona   Warm desert (BWh)
32° 53′ N, 112° 43′ W; elev. 858 ft

	J	F	M	A	M	J	J	A	S	O	N	D	Yr.
Precip. (in)	0.6	0.4	0.6	0.2	0.1	0.1	0.8	1.0	0.4	0.4	0.4	0.7	5.7
Temp. (°F)	53.0	57.1	61.8	69.2	78.2	87.3	94.0	91.8	86.4	74.8	61.8	53.9	

Fog Desert
Walvis Bay, Namibia   Warm desert (BWh)
22° 53′ S, 14° 26′ E; elev. 0 ft

	J	F	M	A	M	J	J	A	S	O	N	D	Yr.
Precip. (in)	0.1	0.1	0.1	0.1	—	—	—	—	—	—	—	—	0.5
Temp. (°F)	64.0	65.0	63.0	61.0	61.0	60.0	58.0	56.0	56.0	57.0	59.0	62.0	

Lima, Peru   Warm desert (BWh)
12° 00′ S, 77° 07′ W; elev. 43 ft

	J	F	M	A	M	J	J	A	S	O	N	D	Yr.
Precip. (in)	0.0	0.0	0.0	0.0	0.0	0.1	0.2	0.1	0.1	0.1	0.0	0.0	0.6
Temp. (°F)	74.0	75.0	74.0	71.0	68.0	65.0	64.0	63.0	63.0	65.0	68.0	71.0	

Cold Desert
Turkestan, Kazahstan   Cold desert (BWk)
43° 16′ N, 68° 13′ E; elev. 686 ft

	J	F	M	A	M	J	J	A	S	O	N	D	Yr.
Precip. (in)	0.9	0.6	1.2	1.1	0.8	0.3	0.5	0.1	0.0	0.6	0.7	1.2	8.0
Temp. (°F)	23.0	30.0	42.0	57.0	68.0	77.0	82.0	78.0	67.0	53.0	38.0	27.0	

Mediterranean Biome

Athens, Greece   Mediterranean (Csa)
37° 54′ N, 23° 44′ E; elev. 69 ft

	J	F	M	A	M	J	J	A	S	O	N	D	Yr.
Precip. (in)	1.9	1.6	1.6	0.9	0.7	0.3	0.2	0.3	0.4	2.1	2.2	2.4	14.8
Temp. (°F)	50.0	50.0	54.0	59.0	67.0	75.0	81.0	81.0	75.0	67.0	59.0	53.0	

San Diego, California   Mediterranean (Csa)
32° 44′ N, 11° 10′ W; elev. 13 ft

	J	F	M	A	M	J	J	A	S	O	N	D	Yr.
Precip. (in)	2.2	1.6	1.9	0.8	0.2	0.1	0.0	0.1	0.2	0.4	1.1	1.4	9.9
Temp. (°F)	57.0	58.0	59.0	62.0	64.0	67.0	71.0	72.0	71.0	67.0	62.0	58.0	

Temperate Grassland Biome

Topeka, Kansas Semiarid (BSk)
39° 04′ N 95° 38′ W; elev. 877 ft

	J	F	M	A	M	J	J	A	S	O	N	D	Yr.
Precip. (in)	1.0	1.3	2.1	3.0	4.4	4.8	3.8	4.1	3.6	2.6	1.6	1.2	33.5
Temp. (°F)	28.6	32.4	42.8	54.9	64.4	73.9	78.8	77.2	69.1	57.6	43.3	32.4	

Amarillo, Texas Semiarid (BSk)
35° 14′ N, 101° 42′ W; elev. 3,590 ft

	J	F	M	A	M	J	J	A	S	O	N	D	Yr.
Precip. (in)	0.5	0.5	0.9	1.1	2.8	3.5	2.7	2.9	2.0	1.4	0.6	0.6	31.6
Temp. (°F)	35.1	39.2	46.9	56.7	65.3	73.9	78.4	76.5	69.1	58.5	45.9	36.9	

Rostov-na-Donau, Russia Semiarid (BSk)
47° 15′ N 39° 49′ E; elev. 252 ft

	J	F	M	A	M	J	J	A	S	O	N	D	Yr.
Precip. (in)	0.7	1.6	1.4	1.6	1.9	2.5	2.1	1.5	1.4	1.5	1.7	2.1	20.9
Temp. (°F)	22.5	23.7	32.9	48.9	61.5	68.5	73.2	71.6	61.2	48.4	36.7	27.9	

Buenos Aires, Argentina Humid subtropical (Cfa)
34° 20′ S 58° 30′ W; elev. 78 ft

	J	F	M	A	M	J	J	A	S	O	N	D	Yr.
Precip. (in)	3.7	3.2	4.6	3.5	3.0	2.5	2.3	2.6	3.1	3.8	3.5	3.8	39.6
Temp. (°F)	74.3	72.9	69.1	62.1	55.9	50.7	50.0	52.0	55.8	60.8	66.7	71.6	

Jan Smuts, South Africa Humid subtropical, dry winter (Cwb)
26° 08′ S, 28° 15′ E; elev. 5,558 ft

	J	F	M	A	M	J	J	A	S	O	N	D	Yr.
Precip. (in)	4.9	3.8	3.3	2.1	0.7	0.3	0.2	0.2	1.1	2.9	4.7	4.3	28.4
Temp. (°F)	66.9	66.2	64.4	59.5	54.5	48.9	49.6	54.3	60.1	63.0	64.2	66.0	

Temperate Broadleaf Deciduous Forest Biome

Roanoke, Virginia   Humid subtropical (Cfa)
37° 19′ N, 79° 58′ W; elev. 1,149 ft

	J	F	M	A	M	J	J	A	S	O	N	D	Yr.
Precip. (in)	2.9	3.2	3.7	3.3	3.9	3.5	3.8	4.0	3.3	3.4	3.0	3.1	40.8
Temp. (°F)	34.3	37.2	46.8	55.6	64.0	71.4	75.6	74.5	67.6	56.5	47.5	38.3	

Wuhan, China   Humid subtropical (Cfa)
30° 37′ N, 114° 08′ E; elev. 75 ft

	J	F	M	A	M	J	J	A	S	O	N	D	Yr.
Precip. (in)	1.6	2.2	3.6	5.3	6.5	8.3	6.5	4.5	2.9	2.9	1.9	1.2	47.4
Temp. (°F)	38.3	41.5	50.4	61.5	71.1	78.3	83.8	83.3	74.5	64.2	52.7	42.4	

Echternach, Luxembourg   Marine west coast (Cfb)
49° 48′ N, 06° 27′ E; elev. 548 ft

	J	F	M	A	M	J	J	A	S	O	N	D	Yr.
Precip. (in)	2.3	2.0	1.7	1.5	1.9	2.7	2.8	2.5	2.2	2.1	2.6	3.0	27.1
Temp. (°F)	34.2	36.5	40.6	45.5	54.0	60.1	62.8	62.1	55.9	47.7	40.1	36.3	

Puerto Montt, Chile (temperate rainforest)   Marine west coast (Cfb)
41° 25′ S, 73° 05′ W; elev. 282 ft

	J	F	M	A	M	J	J	A	S	O	N	D	Yr.
Precip. (in)	4.0	3.8	5.4	6.5	9.5	9.4	9.7	8.5	6.2	5.0	4.9	4.8	78.1
Temp. (°F)	58.3	57.2	54.7	50.7	48.2	44.6	44.2	44.4	46.4	49.6	53.4	56.5	

Davenport, Iowa   Humid continental, hot summer (Dfa)
41° 37′ N, 90° 34′ W; elev. 751 ft

	J	F	M	A	M	J	J	A	S	O	N	D	Yr.
Precip. (in)	1.5	1.3	2.4	3.1	3.8	4.3	3.7	3.5	3.5	2.4	2.0	1.6	33.1
Temp. (°F)	22.1	25.9	36.5	50.2	61.7	71.2	75.9	73.8	65.8	54.3	39.2	27.5	

Boreal Forest Biome

Seattle, Washington   Marine west coast (Cfb)
47° 45′ N, 122° 30′ W; elev. 397 ft

	J	F	M	A	M	J	J	A	S	O	N	D	Yr.
Precip. (in)	5.5	4.2	3.7	2.5	1.7	1.5	0.8	1.1	1.9	3.5	5.9	5.9	38.1
Temp. (°F)	40.1	43.3	45.5	49.1	55.0	60.8	65.1	65.5	60.4	52.7	45.1	40.5	

Quebec City, Quebec, Canada   Humid continental, warm summer (Dfb)
46° 48′ N, 71° 23′ W; elev. 239 ft

	J	F	M	A	M	J	J	A	S	O	N	D	Yr.
Precip. (in)	3.3	2.9	3.1	3.0	3.7	4.3	4.4	4.3	4.4	3.5	3.9	4.1	44.8
Temp. (°F)	18.9	21.7	32.2	46.4	62.1	71.8	77.0	74.1	64.6	52.3	37.8	23.4	

Moscow, Russia   Humid continental, warm summer (Dfb)
55° 58′ N, 37° 25′ E; elev. 623 ft

	J	F	M	A	M	J	J	A	S	O	N	D	Yr.
Precip. (in)	1.4	1.1	1.3	1.5	2.0	2.6	3.2	2.8	2.3	2.0	1.7	1.7	23.6
Temp. (°F)	13.6	16.0	24.8	40.1	54.0	61.3	65.3	61.9	51.6	39.7	28.4	18.5	

Fairbanks, Alaska   Subarctic (Dfc)
64° 49′ N, 147° 52′ W; elev. 436 ft

	J	F	M	A	M	J	J	A	S	O	N	D	Yr.
Precip. (in)	0.6	0.4	0.4	0.3	0.6	1.4	1.9	1.8	1.0	0.8	0.7	0.8	10.7
Temp. (°F)	−9.8	−3.4	11.1	30.7	48.6	59.7	62.4	56.7	45.5	25.2	2.8	−6.2	

Surgut, Russia   Subarctic (Dfc)
61° 25′ N, 73° 50′ E; elev. 200 ft

	J	F	M	A	M	J	J	A	S	O	N	D	Yr.
Precip. (in)	0.9	0.6	0.7	0.9	1.6	2.4	2.8	3.0	2.3	1.7	1.2	1.0	19.0
Temp. (°F)	−7.1	−3.0	9.5	26.2	39.4	55.6	62.8	57.2	45.9	29.3	9.1	−3.2	

Jakutsk, Russia   Humid continental, severe and dry winter (Dwd)
62° 05′ N, 129° 45′ E; elev. 338 ft

	J	F	M	A	M	J	J	A	S	O	N	D	Yr.
Precip. (in)	0.3	0.2	0.2	0.3	0.6	1.2	1.5	1.5	0.9	0.6	0.5	0.3	8.2
Temp. (°F)	−44.6	−33.1	−8.4	18.1	42.1	59.5	66.0	59.0	42.6	17.1	−19.4	−39.5	

Verkhoyansk Russia   Humid continental, severe and dry winter (Dwd)
67° 33′ N, 133° 23′ E; elev. 449 ft

	J	F	M	A	M	J	J	A	S	O	N	D	Yr.
Precip. (in)	0.2	0.2	0.2	0.2	0.3	1.0	1.2	1.1	0.6	0.4	0.4	0.3	6.0
Temp. (°F)	−48.0	−40.0	−18.0	10.0	37.0	55.0	59.0	52.0	37.0	5.0	−31.0	−41.0	

Tundra Biome

Pangnirtung, Nunavut, Canada   Tundra (ET)
66° 01′ N, 65° 43′ W; elev. 49 ft

	J	F	M	A	M	J	J	A	S	O	N	D	Yr.
Precip. (in)	1.0	0.6	0.8	1.0	0.7	1.3	1.4	2.5	2.0	1.9	2.0	1.0	16.2
Temp. (°F)	−14.0	−14.0	−4.0	10.0	26.0	35.0	44.0	42.0	37.0	24.0	12.0	−5.0	

Vardō, Norway   Tundra (maritime variant) (ET)
70° 22′ N, 31° 02′ E; elev. 101 ft

	J	F	M	A	M	J	J	A	S	O	N	D	Yr.
Precip. (in)	2.1	1.6	1.4	1.3	1.3	1.5	1.7	2.0	2.2	2.2	2.0	1.9	21.1
Temp. (°F)	22.8	21.9	24.3	29.8	35.8	42.6	48.0	48.6	43.7	35.4	29.3	25.3	

Temperate Alpine Biome

Niwot Ridge, Colorado   Highland (H)
40° 03′ N, 105° 37′ W; elev. 12,267 ft

	J	F	M	A	M	J	J	A	S	O	N	D	Yr.
Precip. (in)	4.6	3.9	4.6	4.8	3.7	2.3	2.8	2.3	2.0	2.6	4.0	3.6	42.2
Temp. (°F)	9.8	10.0	13.3	20.7	30.8	41.0	47.2	45.3	38.0	27.6	15.7	10.2	

Mount Washington, New Hampshire   Highland (H)
44° 16′ N, 71° 18′ W; elev. 6,626 ft

	J	F	M	A	M	J	J	A	S	O	N	D	Yr.
Precip. (in)	7.1	7.4	7.9	7.2	6.9	7.2	7.0	8.0	7.4	7.1	9.2	8.4	90.7
Temp. (°F)	5.2	5.4	12.4	22.6	34.9	44.2	48.7	47.1	40.5	30.7	20.5	9.0	

Tropical Alpine Biome

Metstation, Mount Kenya   Highland (H)
0° 10′ S, 37° 20′ E; elev. 10,000 ft

	J	F	M	A	M	J	J	A	S	O	N	D	Yr.
Precip. (in)	2.2	3.2	3.9	8.3	6.1	13.2	3.1	3.6	4.7	7.9	7.9	4.5	58.5
Temp. (°F)	54.5	54.9	55.4	55.4	53.6	52.7	52.2	63.6	53.6	54.5	55.4	55.4	

Cuzco, Peru   Highland (H)
13° 33′ S, 71° 59′ W; elev. 10,657 ft

	J	F	M	A	M	J	J	A	S	O	N	D	Yr.
Precip. (in)	5.9	4.5	3.8	1.5	0.3	0.1	0.1	0.3	0.9	1.9	2.7	4.3	26.3
Temp. (°F)	55.6	55.4	55.4	54.5	52.7	50.4	50.0	52.0	54.7	56.5	57.0	56.3	

Source: Various.

Glossary

Adaptation. Any inherited aspect of morphology, physiology, or behavior that improves a species chances of long-term survival and reproductive success in a particular environment.

Adaptive Radiation. Diversification of a species into several new species as different populations develop adaptations to different ecological niches.

Adiabatic Processes. The various steps that result in the lowering or increasing of air temperature without the loss or addition of heat. Adiabatic cooling happens when rising air expands. Adiabatic warming occurs when descending air contracts.

Aerenchyma. Tissue with air-filled cavities found in the roots of plants that grow in water-logged situations such as mangrove swamps. It allows the exchange of gases, including oxygen, between the stem and the roots.

Aerosol. Tiny particles of solids or liquids held in suspension in air.

Anaerobic. Lacking oxygen or functioning in the absence of oxygen.

Angiosperm. A seed-bearing plant that has its seeds enclosed in a protective ovary. See also **Gymnosperm**.

Annual. Pertaining to a plant species that completes its life cycle in one year or less.

Atmospheric Pressure. The force or weight of the atmosphere on a surface. Measured by a barometer and therefore sometimes called barometric pressure.

Autotroph. An organism able to produce its own organic compounds from carbon dioxide. A primary producer.

Benthic. Pertaining to or inhabiting the bottom substrate of a body of water.

Biodiversity. The total variation and variability of life, including not only species but also their genes, the communities and functioning ecosystems in which they live, and the mosaics of landscapes on the Earth.

Biofilm. A slime layer familiar to all who have slipped on river rocks. It consists of bacteria, fungi, periphyton, and other microscopic organisms together with their secretions, in a kind of matrix that adheres to the substrate.

Biome. A geographic region characterized by the dominance of a particular type of vegetation and its associated animals and soils. The lifeforms in the biome are adapted to the climate of the region or some other dominant element of the physical environment, such as edaphic conditions or periodic disturbance. Different taxa may occur in different parts of the same biome.

Bromeliad. Any plant that is a member of the pineapple family (Bromeliaceae). In the humid tropics some so-called tank bromeliads have an inner whorl of leaves fused together to form a water-holding structure. Many are epiphytes, but some such as the pineapple are terrestrial plants.

Bryophyte. A group of multicelled plants that lack vascular tissues to transport water and that reproduce by spores. Mosses, liverworts, and hornworts.

Caatinga. The low dry or seasonal forest and scrublands in northeast Brazil.

Calcification. A major soil-forming process in which the upward movement of soil moisture results in a concentration of calcium carbonate in the B horizon. Associated with semiarid and arid climate regions.

Canopy. The highest layer of foliage in a forest or savanna. It is formed by the crowns of trees or other woody plants. A closed canopy provides 100 percent cover of the ground below. An open canopy allows sunlight to penetrate. See also **Cover**.

Capillary Action. The upward movement of water through confined spaces because of its high surface tension.

Cation Exchange Capacity. The ability of soil to chemically attract positively charged nutrient ions and to release them to plants. A measure of soil fertility. Also referred to as base ion exchange capacity.

Cerrado. The common name of tropical savannas developed on the ancient bedrock of the Brazilian Highlands.

Climate. The general weather patterns expected during an average year. The main factors are temperature and precipitation. See also **Weather**.

Climax (community). A group of plants and animals that represent the final stage of succession. Its structure is influenced by the regional climate and soils and considered more or less stable as long as major changes do not affect the overall environment.

Cnidarian. A member of the phylum Cnidaria, sac-like animals with stinging cells or nematocysts such as coral polyps, jellyfish, and sea anemones.

Community. All the species living in a particular area, or a subset of them such as all the fishes or all the invertebrates or all the animals living in the benthic zone. Some sort of interrelationship among the members of a community is often assumed.

Conifer. A cone-bearing tree such as pine or spruce. More formally, a woody plant of the Division Pinophyta or Coniferae. They are the most common among the four groups called gymnosperms.

Continentality. The temperature pattern typical of high- and mid-latitude locations at some distance from the sea. Wide temperature ranges between summer and winter are characteristic.

Continental Shield. Exposed ancient (Precambrian) bedrock or basement rock of the continents, usually granitic.

Convection. The upward movement of fluids induced by warming and lowering the density of air, water, or mantle at lower levels. Warm fluids rise through cooler, denser fluids above them.

Convergent Evolution. The development of similar morphological or other characteristics in unrelated taxa under similar environmental conditions.

Coriolis Force. The apparent deflection of moving fluids and objects because of the rotation of the Earth. In the Northern Hemisphere, the deflection is to the right of the intended path; in the Southern Hemisphere, it is to the left.

Cover. The area of the ground overlaid or shaded by a particular plant species or layer of the vegetation. Usually expressed as a percent.

Crepuscular. Active at dawn and dusk.

Cuticle. A waxy coating on leaves that lowers water loss from leaf cells.

Cyanobacteria. Bacteria that live in water and soil and are able to photosynthesize and fix nitrogen. Also called blue-green algae.

DDT (dichloro-diphenyl-trichloroethane). A persistent organic pesticide first manufactured and used in large quantities during World War II to control mosquitoes and other disease-carrying insects. Later it was widely applied as an insecticide in agriculture. DDT proved toxic to a wide range of invertebrates and vertebrates and became a major threat to birds because it led to thinning of the eggshell. Implicated in the near extinction of the Bald Eagle, Peregrine Falcon, and Brown Pelican, DDT was banned in the United States in 1972, and populations of these birds began to recover. It was banned worldwide in 2004, although limited use is still permitted to fight malaria.

Decomposer. An organism that breaks down dead organic material into simpler molecules or its inorganic components.

Density. The number of some object per unit area or volume.

Deposit-feeders. Aquatic organisms that feed by ingesting organic particles in or on the bottom of a body of water.

Detritivore. A consumer that feeds on decomposing organic material (detritus).

Detritus Food Chain. The path through which energy originally held in waste products and dead organic matter passes through detritivores and decomposers in an ecosystem. See also **Grazing Food Chain**.

Dew Point. The temperature to which air must be lowered for it to become saturated.

Dispersal. In an ecological sense, the movement of spores, seeds, or offspring away from the parent individual. In a biogeographic sense, the movement of a species beyond its present range and its establishment in new territory.

Ecology. The interrelationships among organisms and the living and nonliving aspects of their environment; the science that studies these interrelationships.

Ecosystem. All the living and nonliving parts of a given area that work together as a single unit to maintain a flow of energy and cycling of nutrients.

Edaphic. Pertaining to conditions of the soil or substrate, especially those conditions that limit the growth of plants. Edaphic factors such as waterlogged soils, hardpans, or nutrient deficiencies may determine the nature of the vegetation occupying certain sites.

Endemic. Native to and restricted to a particular geographic area.

Ephemeral. Pertaining to a plant species that completes its life cycle in a few weeks.

Epiphyte. A plant that grows on other plants with no contact with the ground and obtains water and nutrients from the air rather than from the host or soil.

Equinox. A position on the Earth's orbit where, because of the orientation of Earth relative to the sun, all points on the planet receive 12 hours of daylight and 12 hours of darkness. Such positions on the orbit are reached in March and September. See also **Solstice**.

Evapotranspiration. The movement of water vapor from plants and soil to the atmosphere through a combination of transpiration and evaporation.

Evolution. Changes in the genetically controlled traits of a population or species from generation to generation.

Exposure or Aspect. The direction toward which a slope faces.

Fauna. All the animal species in a given area, or some subset of them such as the bird fauna or grazing fauna.

Feedback. A response of one part of a self-regulating system to changes elsewhere in the system that sends a signal or response back to the source of change to control the system. The path of the signal back to the origin of change is a feedback loop. Feedback is negative when the signal adjusts the system back to its original equilibrium state. Feedback is positive when it allows or forces the system to change to a new state. Negative feedback loops are significant in regulating predator-prey population dynamics. Positive feedbacks are anticipated by the excess of carbon dioxide and other greenhouses gases being put into the atmosphere and forcing global climate change.

Filter-feeders. Organisms than strain plankton and other organic particles from the water.

Flora. A collective term for all the plant species occurring in a given area.

Forb. An herb with broad leaves and soft, nonwoody stem. Wildflowers are typical of this growthform.

Fynbos. The local name for mediterranean scrub vegetation in South Africa's Cape Floristic Province.

Gallery Forest. The strip of trees lining a river in a savanna or other nonforested type of vegetation.

Graminoid. The growthform of grasses, sedges, rushes, and reeds. A type of herb.

Grazer. On land, a plant-eating animal that consumes primarily grasses. In aquatic ecosystems, an animal that consumes algae, cyanobacteria, or macrophytes.

Grazing Food Chain. The path of solar energy fixed by green plants through the living herbivores and carnivores of an ecosystem. See also **Detritus Food Chain**.

Growthform. The size, shape, and other traits that provide a distinctive appearance to basic types of plants such as trees, shrubs, herbs, and succulents.

Gymnosperm. Seed-bearing plants with naked seeds. The seeds are usually held on the scales of a cone or cone-like structure. See also **Angiosperm**.

Habitat. The space in which a species lives and the environmental conditions of that place.

Herb. A nonwoody plant that dies down each year. May be an annual or a perennial. This growthform includes graminoids and forbs.

Herbaceous. Having the characteristics of an herb.

Heterotroph. An organism that must depend on other species to fix the energy and carbon it uses.

Humus. Well-decayed plant matter that is colloidal in size and assumes a dark-brown color. Since it helps hold moisture and nutrients in a soil, humus content is an indicator of soil fertility.

Inorganic. Pertaining to nonbiologically produced compounds.

Isotope. A variation of an element with a different atomic number. All isotopes of the same element have the same number of protons in their nuclei but different numbers of neutrons.

ITCZ (Intertropical Convergence Zone). The contact zone between the Trade Winds of the Northern and Southern hemispheres. Shifts its position north and south of the Equator with the seasons and, when overhead, usually brings rain.

Katabatic. Pertaining to winds that are generated when gravity draws high-density air downslope. Most often associated with continental ice sheets.

Laterization. A soil-forming process in which silica is mobilized and removed from the A horizon along with soluble bases. Aluminum and iron remain in the A horizon and give the resulting soils a deep red or yellow color. Associated with humid tropical climates.

Latitude. The distance of a point north or south of the Equator (0° Latitude), measured in degrees.

Leaching. The process in which dissolved substances are removed from the substrate by the downward percolation of water.

Liana. A woody vine.

Life Zone. A belt of characteristic plants or animals reflecting the environmental differences that occur at certain elevations and distinct from surrounding zones. May refer only to those altitudinal zones replicated in broad latitudinal belts in the mid-latitudes, as in Merriam's Life Zones.

Litter. The dead leaves, only partially decayed and still recognizable as parts of a plant, and any other organic debris that accumulates on the ground surface.

Littoral. Pertaining to the shore of a body of water.

Loess. Powder-like silts transported and deposited by the wind. The most fertile temperate soils, the chernozems, are generally developed on this type of substrate.

Lotic. Pertaining to aquatic habitats in flowing water, as in a stream or river.

Macrophyte. A plant large enough to be seen with the unaided eye.

Mantle. The middle layer of planet Earth lying between the crust and core. Part of this layer is plastic and moves in slow convection currents that drive plate tectonics.

Maritime Influence. The moderating affect of a large body of water on air temperature.

Mediterranean. Refers to regions or climate patterns where winter is the rainy season and summers are dry.

Microbial Loop. The food chain in which dissolved organic matter (DOM) is leaked from algal and zooplankter cells and consumed by marine bacteria, which then are eaten by zooplankters.

Monsoon. A wind that reverses its direction seasonally. An onshore flow typifies the warm season and an offshore flow occurs during the cold season. The Asian Monsoon is most powerful and dominates the climate of the vast Indian Ocean region. More localized monsoonal systems occur elsewhere, as in the American Southwest.

Morphology. The form (shape and size) and structure of an organism.

Neotropical. Pertaining to the region from southern Mexico and the Caribbean to southern South America or to animals and plants restricted or nearly restricted to this region.

Niche. The ecological role of a species in an ecosystem. How a species obtains energy and nutrients, or its position along the flow of energy through the system.

Ombrotrophic (wetlands). Wetlands fed exclusively or mainly by rainwater, which is typically very low in certain key plant nutrients.

Organic. Pertaining to complex compounds of carbon produced by living organisms.

Orographic Effect. The increase in relative humidity and precipitation caused by the adiabatic cooling of air forced to rise over a mountain range or other land barrier.

Parasite. An organism that obtains its energy and nutrients by tapping into the supplies provided by a host organism.

Parent Material. The solid rock or unconsolidated particles that weather and form the mineral component of soils.

PCBs (polychlorinated biphenyls). A group of persistent organic pollutants formerly in widespread industrial use as coolants, insulating fluids in transformers and capacitors, and additives to polyvinyl chloride products such as pipes, vinyl siding, and phonograph records. They were determined to cause cancer and be toxic to animals, negatively affecting the liver and immune system and disrupting the endocrine system. They have not been produced in the United States since 1977 but are still found in the environment.

Perennial. A plant that lives two or more years.

Periphyton. The community of algae, cyanobacteria, and other microorganisms, combined with detritus, that develops on submerged surfaces.

pH. A measure of acidity (0–7) or alkalinity (7–14). The negative logarithm of the concentration of hydrogen ions in solution.

Phenology. The timing of biological or life history events, such as flowering or green-up or hatching of eggs.

Phylogeny. The evolutionary history of a taxon.

Physiognomy. The characteristic appearance of a plant community.

Phytoplankton. A collective term for all plants that float in the water unable to move against tides and currents. Many, however, can propel themselves up and down the water column. The individual cells or organisms are **phytoplankters**.

Plankter. An individual cell or small organism that floats in the currents or tides unable to change location by itself, except up or down in the water column.

Plankton. A collective term for all organisms that float in the water unable to move against tides and currents.

Plate Tectonics. The movement of pieces of the Earth's crust (plates) and the rearrangements and deformation of the surface that result.

Pneumatophore. Specialized roots that allow oxygen to pass from the air to the roots in waterlogged soils. They may grow up from regular roots or down from the stems.

Podzolization. A soil-forming process that occurs under acidic conditions. Most compounds other than silica oxide are mobilized and leached from the A horizon, producing a distinct whitish or ashy layer. Materials removed from the A horizon may accumulate in the B horizon. Most directly associated with the needleleaf boreal forests, podzolization occurs under pines in any climate region.

Polar Front. In the global atmospheric circulation system, the contact zone between cold polar air and warmer subtropical airmasses. Associated with uplift at the subpolar lows and the often stormy, wet weather patterns of the mid-latitudes.

Population. A subset or group of individuals of a single species limited to a specific geographic area and usually isolated from other populations.

Potential Evapotranspiration. The maximum amount of moisture that could be evaporated from an area if the soils were saturated. In humid regions, precipitation is greater than potential evapotranspiration. In semiarid and arid climates, potential evapotranspiration is greater than actual precipitation.

Primary Production. The amount of energy and carbon fixed by green plants in a given area.

Primary Productivity. The rate at which green plants fix energy and carbon by photosynthesis. Given in weight of new biomass per unit area per unit time.

Primary Succession. The development of a series of plant communities beginning with a new site devoid of soil or plants. See also **Succession**.

Rainshadow. A dry region that develops on the lee (downwind) side of mountain ranges. Low amounts of precipitation are the result of the warming of airmasses as they descend the mountain slopes.

Relative Humidity. A ratio comparing the amount of water vapor actually in an airmass with the total amount that could be held at the current air temperature. Expressed as a percent.

Rendzina. A dark soil that usually develops on limestone or other carbonate bedrock. Common in mediterranean climate regions.

Riparian. Along a stream's margins. Used to refer to the moist habitats or vegetation alongside flowing waters in any biome or climate region.

Rosette. A growthform in which the leaves of a plant radiate from a central point and overlap like the petals of a rose.

Salinity. A measure of the amount of dissolved salts in water.

Saprotroph. Any organism that obtains its energy and nutrients from dead organic matter.

Scale. The relationship between the distance or size on an area displayed on a map and the true dimensions of the area on Earth. Indicates the degree of detail or of generalization that can be represented or visualized.

Scavenger. An animal that feeds on carrion (dead animals).

Secondary Succession. The development of a series of plant communities beginning on a site that has a well-developed soil but from which the vegetation has been cleared. See also **Succession**.

Sere. A series of plant communities developed during ecological succession. A sere that appears on a site that is initially solid bedrock will differ from one that begins on open water. Only the final community will be the same, as determined by the regional climate.

Solstice. A position on the Earth's orbit where, because of the planet's position relative to the sun, one pole will experience 24 hours of total darkness and the opposite poles will experience 24 hours of daylight. Such positions are reached twice a year, in June and in December. See also **Equinox**.

Species. A group of individual organisms that can interbreed and produce viable offspring.

Species Diversity. The number of taxa found in a specified geographic area or community adjusted for the relative abundance of each. Several mathematical formulas have been devised to express species diversity. Also used as synonym for **species richness**.

Species Richness. A count of the number of species found in a specified geographic area or community.

Strangler. A plant of humid tropical environments that as a seedling is an epiphyte in the tree canopy, but that sends its roots down to the ground and, as a mature plant, obtains energy and nutrients from the soil. A network of thick roots may come to envelop the host tree, which eventually dies.

Stratovolcano. A large cone-shaped volcano composed of alternating layers of ash and lava that form on continents above subduction zones.

Subclimax. A persistent plant community that does not reflect the regional climate (as a climax community does), but rather is maintained long term by edaphic or disturbance factors.

Subsoil. The B horizon of a soil.

Subsolar Point. That place on the Earth where the noon sun strikes at a 90-degree angle.

Substrate. Soil, sediment, bedrock, or other surface materials. Also the parent material for soil development.

Succession (ecological). The orderly and predictable process in an ecosystem whereby one community of plants and animals is replaced by another, culminating in a stable "climax" community, which exists for a particular set of climatic and geomorphic conditions. A somewhat outdated concept. See also **Primary Succession; Secondary Succession**.

Suspension-feeders. Aquatic organisms that sieve food particles from the water. See also **Filter-feeders**.

Taxon (plural = taxa). Any group at any level in the taxonomic hierarchy.

Taxonomy. The way scientists have classified a group of similar, related organisms into species, genera, families, and higher units. Also, the science that classifies, describes, and names organisms.

Topsoil. The A horizon, especially if rich in humus.

Tropics. The latitudinal zone on Earth that lies between 23° 30′ N and 23° 30′ S—that is, between the Tropic of Cancer and the Tropic of Capricorn.

Turbulence. Chaotic motion in the flow of fluids. Smooth directional flow is called laminar flow.

Varve. An annual layer of sediment deposited in a lake bed.

Vegetation. The general plant cover of an area described in terms of its structure and appearance and not the species that comprise it.

Vegetation Structure. The vertical arrangement of foliage and horizontal spacing of plants that reflects the dominant growthforms present and the regional climate.

Water Column. In a body of water, a vertical segment extending from the surface to the bottom.

Weather. The state of the atmosphere at any given moment. Includes atmospheric pressure, temperature, humidity, and type of precipitation (if any).

Weathered. Pertaining to bedrock that has undergone physical and/or chemical breakdown into small particles, even ions.

Weed. A plant adapted to invade disturbed sites. Generally short-lived, weeds are good dispersers and fast growers.

Zonal Soil. A soil whose characteristics are largely determined by the climate type and its associated vegetation. Since latitude is a major control of climate, these soils show a general latitudinal pattern or zonation in their distribution.

Zonation. A distribution pattern in which particular forms of life occur in distinct belts.

Zooplankton. A collective term for all the single-celled and small multicelled animals that float in the ocean unable to move against tides or currents but able to move up and down the water column.

Bibliography

Atmosphere, Climate, and Environment Information Programme. 2002–2004. "Global Climate Change Student Guide." Manchester Metropolitan University. http://www.ace.mmu.ac.uk/resources/gcc/contents.html.

Bailey, Robert G. 1996. *Ecosystem Geography*. New York: Springer-Verlag.

Bailey, Robert G. 1998. *Ecoregions. The Ecosystem Geography of the Oceans and Continents.* New York: Springer-Verlag.

Bakun, A., and S. Weeks. 2004. "Greenhouse Gas Buildup, Sardines, Submarine Eruptions and the Possibility of Abrupt Degradation of Intense Marine Upwelling Ecosystems." *Ecology Letters* 7: 1015–1023.

Brown, James H., and Mark V. Lomolino. 1998. *Biogeography*, 2nd ed. Sunderland, MA: Sinauer Associates, Inc. Publishers.

Canty & Associates. 2008. Weatherbase. http://weatherbase.com.

Cole, Kenneth. 2002. "Packrat Middens." In *Canyons, Cultures and Environmental Change: An Introduction to the Land-Use History of the Colorado Plateau*, ed. John D. Grahame and Thomas D. Sisk. http://www.cpluhna.nau.edu/Tools/packrat_middens.htm.

CP-LUHNA. n.d. "Fossil Dung." In *Canyons, Cultures and Environmental Change: An Introduction to the Land-Use History of the Colorado Plateau*, ed. John D. Grahame and Thomas D. Sisk. http://www.cpluhna.nau.edu/Tools/fossil_dung.htm.

Dansereau, Pierre. 1957. *Biogeography. An Ecological Perspective*. New York: The Ronald Press Company.

Donlan, C. J., J. Berger, C. E. Bock, J. H. Bock, D. A. Burney, J. A . Estes, D. Foreman, P. S. Martin, G. W. Roemer, F. A. Smith, M. E. Soulé, and H. W. Greene. 2006. "Pleistocene Rewilding: An Optimistic Agenda for Twenty-First Century Conservation." *The American Naturalist* 168: 660–681.

Earth Observatory. n.d. "What Is a Coccolithophore?" NASA. http://earthobservatory
.nasa.gov/Library/Coccolithophore.

Eyre, S. R. 1968. *Vegetation and Soils*, 2nd ed. Chicago: Aldine Publishing Company.

Furley, P. A., and W. W. Newey. 1983. *Geography of the Biosphere. An Introduction to the Nature, Distribution and Evolution of the World's Life Zones*. London: Butterworths.

Gleason, H. A. 1926. "The Individualistic Concept of the Plant Association." *Torrey Botanical Club Bulletin* 53: 7–26.

Good, Ronald. 1947. *The Geography of Flowering Plants*. London: Longmans, Green and Company.

Hoare, Robert. 1996–2008. WorldClimate. http://www.worldclimate.com.

Holzman, Barbara A. 2008. *Tropical Forest Biomes*. Greenwood Guides to Biomes of the World. Westport, CT: Greenwood Press.

Klinkenberg, Brian, ed. 2007. E-Flora BC: Electronic Atlas of the Plants of British Columbia [www.eflora.bc.ca]. Lab for Advanced Spatial Analysis, Department of Geography, University of British Columbia, Vancouver.

Köppen, Wladimir. 1923. *Grundiss der Klimakinde: Die Climate die Erde*. Berlin: DeGruyter.

Kuennecke, Bernd H. 2008. *Temperate Forest Biomes*. Greenwood Guides to Biomes of the World. Westport, CT: Greenwood Press.

Lomolino, Mark V., Dov F. Sax, and James H. Brown. 2004. *Foundations of Biogeography. Classic Papers with Commentaries*. Chicago: University of Chicago Press.

MacDonald, Glen. 2003. *Biogeography. Introduction to Space, Time and Life*. New York: John Wiley & Sons.

McKnight, Tom L., and Darrell Hess. 2000. *Physical Geography. A Landscape Appreciation*, 6th ed. Upper Saddle River, NJ: Prentice Hall.

Moore, Charles. 2003. "Trashed: Across the Pacific Ocean, Plastics, Plastics, Everywhere." *Natural History* 112: 46–53.

Odum, Eugene P. 1959. *Fundamentals of Ecology*, 2nd ed. Philadelphia: W. B. Saunders.

Parmesan, Camille. 2006. "Ecological and Evolutionary Responses to Recent Climate Change." *Annual Review of Ecology, Evolution, and Systematics* 37: 637–669.

Quinn, Joyce A. 2008. *Arctic and Alpine Biomes*. Greenwood Guides to Biomes of the World. Westport, CT: Greenwood Press.

Quinn, Joyce A. 2008. *Desert Biomes*. Greenwood Guides to Biomes of the World. Westport, CT: Greenwood Press.

Real, Leslie A., and James H. Brown. 1991. *Foundations of Ecology. Classic Papers with Commentaries*. Chicago: University of Chicago Press.

Roth, Richard A. 2008. *Freshwater Aquatic Biomes*. Greenwood Guides to Biomes of the World. Westport, CT: Greenwood Press.

Smith, Charles H. 2002–2008. "The Alfred Russel Wallace Page." Western Kentucky University. http://www.wku.edu/%7Esmithch/index1.htm.

Smith, Charles H. 2002–2008. "Early Classics in Biogeography, Distribution, and Diversity Studies: To 1950." Western Kentucky University. http://www.wku.edu/~smithch/biogeog/#L.

Smith, Charles H. 2003–2008. "Early Classics in Biogeography, Distribution, and Diversity Studies: 1951–1975." Western Kentucky University. http://www.wku.edu/%7Esmithch/biogeog/index2.htm.

Smith, Charles H., Joshua Woleben, and Carubie Rodgers. 2002–2008. "Some Biogeographers, Evolutionists and Ecologists: Chrono-Biographical Sketches." Western Kentucky University. http://www.wku.edu/%7Esmithch/chronob/homelist.htm.

Smith, Robert Leo. 1992. *Elements of Ecology*. New York: HarperCollins Publishers, Inc.

Torsvik, Trond H. 2003. "The Rodinia Jigsaw Puzzle." *Science* 300: 1379–1381.

Waggoner, Ben. 1996, 2000. "Carl Linneaus (1707–1778)." University of California Museum of Paleontology. http://www.ucmp.berkeley.edu/history/linnaeus.html.

Williams, John W., and Stephen T. Jackson. 2007. "Novel Climates, No-Analog Communities, and Ecological Surprises." *Frontiers in Ecology* 5: 475–482.

Williams, John W., Stephen T. Jackson, and John E. Kutzbach. 2007. "Projected Distributions of Novel and Disappearing Climates by 2100 AD." *Proceedings of the National Academy of Sciences* 104: 5738–5742.

Woodward, Susan L. 1996. "Introduction to Biomes." Radford University: http://www.runet.edu/~swoodwar/CLASSES/GEOG235/biomes/intro.htm.

Woodward, Susan L. 2003. *Biomes of Earth: Terrestrial, Aquatic, and Human-Dominated*. Westport, CT: Greenwood Press.

Woodward, Susan L. 2008. *Grassland Biomes*. Greenwood Guides to Biomes of the World. Westport, CT: Greenwood Press.

Woodward, Susan L. 2008. *Marine Biomes*. Greenwood Guides to Biomes of the World. Westport, CT: Greenwood Press.

Index

About the Author

SUSAN L. WOODWARD received her Ph.D. in geography from the University of California–Los Angeles, in 1976. She taught undergraduate courses in biogeography and physical geography for twenty-two years at Radford University in Virginia, before retiring in 2006. Author of *Biomes of Earth,* published by Greenwood Press in 2003, she continues to learn and write about our natural environment. Her travels have allowed her to see firsthand some of the world's major terrestrial biomes and at least glimpse marine habitats in the United States, Russia, China, Brazil, Chile, Ecuador, Peru, South Africa, and Namibia.